Contents

Introduction
1. Article 1040. Aliens have filled humanity's knowledge base with lies
2. Article 1105. Earth's radius increased by 82% during the Flood cataclysm due to Planet X
3. Article 1106. Moon's tidal pattern and the size of Earth's central core: 1745 km
4. Article 1011. Planet X cataclysm: size and closeness of the objects responsible
5. Article 1015. In search of Vulcan: Planet X observed traversing the Sun since 1818
6. Article 1028. Earth's new moons and size of Planet X objects observed in 1800s
7. Article 1030. Mercury is not a planet: it is a part of the Planet X System
8. Article 1058. The Moon is a Planet X planetary central core and not a spaceship
9. Article 1093. Space junk in orbit around the earth: Planet X cover-up
10. Article 1110. Planet X in the sky over Northern California: gravity connections
11. Article 1111. Planet X orbits and the Moon creates its own atmosphere
12. Article 1112. The earth's atmosphere creates daylight on one side: the reason
13. Article 1049. Electrons are photons moving at less than the speed of light
14. Article 655. The destruction of the Planet X planets and the electron
15. Article 645. Moon in atmosphere due to Planet X and electrons are light
16. Article 1050. The wave nature of particles comes from the gravitational interaction
17. Article 1051. Earth emits cosmic radiation at bow shock due to Planet X
18. Article 818. Planet X observations: the electrical universe is made of light
19. Article 779. Meteors are Planet X debris: glow is due to electric current
20. Article 1166. Gravitational potential and the size of earth's central core
21. Article 1167. How could astronomers have missed this? Gas giants tell us G is not constant
22. Article 1171. How did the Planet X planets break into pieces?
23. Article 1173. Determining Venus' true mass with the help of Planet X
24. Article 1180. The photon model: all isolated objects have an aura
25. Article 1182. Electrical density and electrogravitic drives
26. Article 1184. Static electricity on human body generated by the human creative spirit
27. Article 620. Huge gas giant planets in the inner Solar System
28. Article 621. Gas Giant vindication and why Dr Eugene Shoemaker was really killed
29. Article 103. The 800 million solar masses black hole at the edge of the universe
30. Article 126. White Holes instead of Black Holes at the Center of Galaxies

Books previously published

Book 1: Planet X: the awakening is now.
Book 2: The Planet X Report 2017: Photographic Evidence.
Book 3: Planet X Revealed Gravity and Light.
Book 4: The Sun Simulator
Book 5: Chemtrails: The Silent Killer.
Book 6: Planet X Physicist Articles: Part 1
Book 7: Planet X: The effects on the Earth and the Sun
Book 8: Planet X and the Solar System
Book 9: Planet X and the Hurricane Michael Cover-Up
Book 10: Planet X Reveals How the Universe Works
Book 11: Planet X and the Bible
Book 12: Planet X: the Greatest Cover Up in World History
Book 13: Planet X Reveals that the Universe is alive

Introduction

Planet X observations lead to the discovery that most of the physics that is being taught on this planet and almost everything of a scientific nature in humanity's knowledge base is a lie or based on lies. This book is based on the many articles that I have written which expose the lies and those in which I attempt to model the true laws of the universe based on the observational evidence that has come across my path. In order to create a theoretical framework which makes sense of the observations it is necessary to resort to the use of some mathematics but nothing terribly complex is necessary, anyone can understand the true physics of the universe. There are already many articles in which I use pure deductive reasoning starting from the observations to reach a conclusion or deeper understanding of the universe. I am grateful to God that my initial desire to understand the strange colors in the sky and what was happening in the Solar System has been satisfied to the point that it can now be described in a simple mathematical form and that the lies, which fill humanity's knowledge base have become very obvious due to their illogical nature.

In Article 1040: Aliens have filled humanity's knowledge base with lies, which appears in chapter 1, I detailed some of the obvious and illogical lies that the prime alien and enemy of mankind has filled the scientific knowledge on this planet with. These lies lead to a diagrammatically opposite view of the real universe and how it functions. Planet X research shows that all celestial objects are created from the inside outwards by core systems, which generate gravity, and that gravity is a creative force, as it leads to the creation of matter. However, according to the scientific understanding on this planet, gravity pulls things inwards and planets therefore form as a result of matter accretion, in other words, pieces of rock or dust come together to form a planet. This is easily shown to be impossible, because the impulsive forces, when any piece of dust or rock collide, even at very tiny speeds, are many orders of magnitude stronger than the gravitational attraction, between the pieces, so this theory is illogical and the fact that scientist all around the planet accept it, unquestionably, shows to what degree these aliens have controlled and blinded their minds.

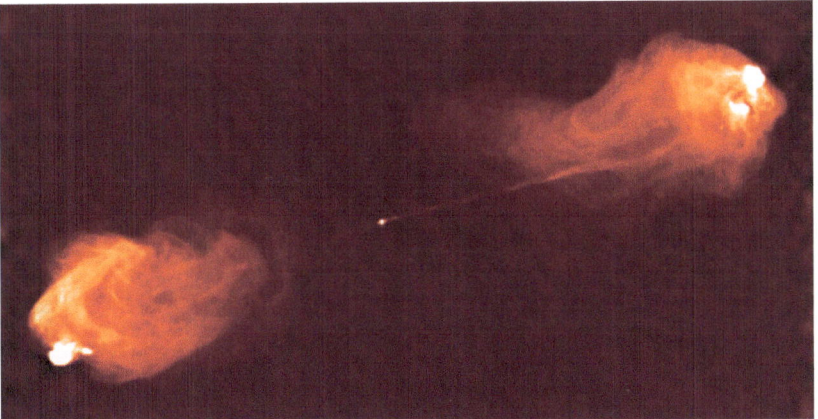

Figure 1. Cygnus A, is a galaxy, which ejects high energy radio emitting material from its nucleus. Galaxies eject matter from their nuclei because there is a source of gravity, a galactic core inside them which creates matter. This is what the astronomer Halton Arp found and detailed in his book: Seeing

Red, which also clearly details irrefutable observational evidence showing that redshift is an intrinsic property of matter and that it cannot be used to determine distance and rate of expansion of the universe and thus falsifies the Big Bang model (see Article 126: White Holes instead of Black Holes at the Center of Galaxies).

Dr. Claudia Albers
Planet X Physicist
October 13th 2019

Chapter 1

1040. Aliens have filled humanity's knowledge base with lies

Planet X is the core systems left over from the destruction of the Planet X planets. These cores are coming to the earth accompanied by the debris fields, which are the broken pieces of these planets, and they have been coming in for thousands of years. These cores are creating cloud, rain and lightning in the earth's atmosphere and yet no scientist, on earth, seems to have discovered the truth, about this, in the last few thousand years (see Article 785: Planet X is here but what is it exactly?) [1] How is this possible? Planet X debris is bringing in insect eggs that are leading to insect plagues and yet no one seems to have discovered that the plagues are coming in from space. Instead, the illogical explanation that insects can suddenly multiply from nothing to billions, overnight, or that a hoping insect can fly over the ocean, to get to a certain location, where they have never appeared before is completely acceptable.

Figure 1.1. Millions of locusts eat all there is to eat on the island of Sardinia, an island belonging to Italy in the Mediterranean Sea in 2019, driving local farmers to desperation and yet there have been no plagues like this on this island in remembered history. How is this possible? How can intelligent scientists believe such illogical explanations?

The theory of dust accretion for planetary formation was proposed in 1944 and the theory is very easily shown to be completely impossible with one simple calculation, so how come astrophysicists still use it and believe its correct today?

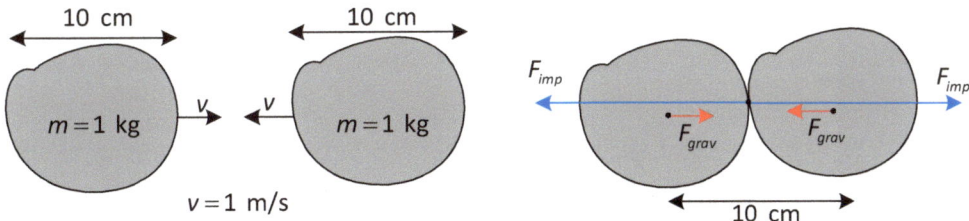

Figure 1.2. Collision of two pieces of rock: solid rock is not a deformable material and therefore the collision is expected to be elastic so that the 2 pieces will fly apart as a result of the impulsive forces.

Let us suppose that the collision time interval is 0.1 s. Then, the impulse force is given by

$$F_{imp} = \frac{2mv}{\Delta t} = \frac{2(1 \text{ kg})(1 \text{ m/s})}{0.1 \text{ s}} = 20 \text{ N}$$

This is the force that will drive the 2 pieces of rock apart after the collision. Now, the force exerted by gravity on the 2 pieces of rock at the point of closest approach is given by

$$F_{grav} = \frac{Gm^2}{r^2} = \frac{(6.67 \times 10^{-11} \text{ Nm}^2/\text{kg}^2)(1 \text{ kg})^2}{(0.1 \text{ m})^2} = 6.67 \times 10^{-9} \text{ N}$$

This is the force that would tend to cause the two pieces of rock to stick together and it is an extremely small force. We can thus see that the impulse force, which will tend to drive the two pieces of rock apart, is 3 x 10⁹ or 3 billion times stronger. And this is with rocks moving at extremely low speeds, with respect to each other, which makes the impossibility of the accretion theory quite obvious. So how can intelligent astrophysicist believe in such an illogical and impossible theory?

The explanation for the tides is the third unacceptable theory, which is believed and taught on earth whilst being clearly impossible. Spring tides supposedly produce ocean levels up to 30% higher or lower than normal tides and this difference is supposedly provided by the fact that the Sun and the moon are aligned but a very simple calculation clearly shows that the Sun's tidal effect is only 0.25% of the moon's so how can the Sun provide the necessary force to provide a 30% higher tide? It is clearly impossible.

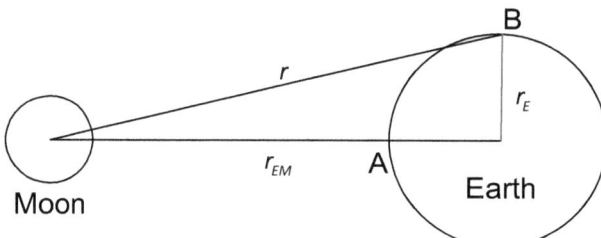

Figure 1.3. To calculate the moon's tidal effect we determine how much stronger the moon's gravitational attraction is at A than at B. The distance between the center of the moon and point A is $r_A = r_{EM} - r_E$ and the distance between the center of the moon and point B is given by $r = r_B = \sqrt{r_{EM}^2 + r_E^2}$.

Then, the difference in the gravitational attraction by the moon between points A and B is given by

$$\Delta F_{GM} = F_{GMA} - F_{GMB} = \frac{GM_M M_E}{r_A^2} - \frac{GM_M M_E}{r_B^2}$$

But we actually only need the relative difference, which is given by

$$\frac{\Delta F_{GM}}{F_{GMA}} = 1 - \frac{F_{GMB}}{F_{GMA}} = 1 - \frac{r_A^2}{r_B^2} = 1 - \frac{(r_{EM} - r_E)^2}{r_{EM}^2 + r_E^2} = 0.033$$

Since the distance between the center of the moon and the center of the earth is $r_{EM} = 3.844 \times 10^5$ km and the radius of the earth is $r_E = 6371$ km. Then, if we repeat the calculation for the Sun, we get:

$$\frac{\Delta F_{GS}}{F_{GSA}} = 1 - \frac{F_{GSB}}{F_{GSA}} = 1 - \frac{r_A^2}{r_B^2} = 1 - \frac{(r_{ES} - r_E)^2}{r_{ES}^2 + r_E^2} = 0.000084$$

So the difference in gravitational attraction which the Sun can provide which is what leads to tidal effects is much weaker than the moon's, it is 0.000084/0.033 = 0.0025 that of the Moon, i.e. the Sun's tidal force is only 0.25% that of the moon's, which clearly shows that the sun's tidal effect cannot in any way provide a 30% increase in the ocean level. It is impossible.

We now therefore have 3 theories, which are accepted by all scientists on earth and all 3 are impossible. And this is only the beginning as the Big Bang theory has been falsified by Halton Arp's work and yet it continues to be used as if nothing had happened. Evolution is illogical and circular arguments regarding strata and age of fossil records, are used; there are no examples of any intermediary species and even believing that animals normally become fossils is impossible, as then fossils would be forming all the time but examination of the ocean floor shows that no such thing happens as carcasses are consumed by creatures at the bottom of the ocean. Clearly the fossil record was produced at one time, at the flood, when huge numbers of species died in a short amount of time and were buried in mud. Then there is the impossible theoretical practices used in particle physics, things like renormalization, where one infinity is subtracted from another, and somehow we are supposed to get something, which makes sense and tells us more about the universe? If scientists are incapable of simple logical processes how can they suddenly expect to understand the universe with such an illogical type of theoretical approach? The obtaining of infinities in calculations should tell theoreticians one thing and one thing only: the theory is wrong.

But how can intelligent scientists believe such illogical theories and refuse to consider what is there for anyone to see: Planet X core matter is inside the earth's atmosphere, the earth has a core system in it that creates matter, matter is being created all over the universe? They seem to be asleep, their minds being manipulated to believe impossible and illogical lies, and being blocked from ever perceiving the truth. Who would do this to humanity's scientists? Obviously, aliens who hate us and want to hide the truth from us. And they have been at it for a very long time.

Sun halos have been observed for at least 100s of years but yet the official explanation that it is produced by ice crystals in the atmosphere is once again illogical and impossible. Sun halos can only be produced by an artificial lens in the sky, which would be a part of a sun simulation device.

Figure 1.4. Image showing a painting from 1535 when several Sun halos appeared in the skies (see Article 545: Planet X and Sun Halos: aliens covering up Sun going dark) [4].

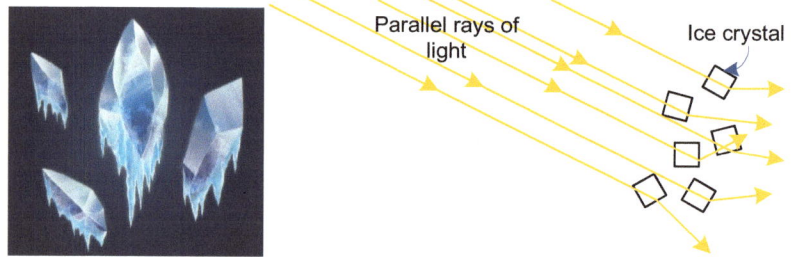

Figure 1.5. It is impossible for ice crystals to be perfectly aligned so that light comes from each crystal at exactly the same angle, thus the result is the same as light being scattered by droplets of water, making up a cloud, which causes clouds to simply appear to be white. Thus, ice crystals in the sky can only look like a cloud. The production of a perfect circular halo is impossible.

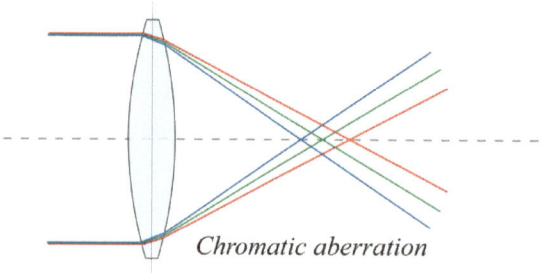

Figure 1.6. Chromatic aberration is the splitting of colors at the edge of a lens. When light passes through the edge of a lens different colors of light are refracted by different amounts, which splits the light into its constituent colors, thus, creating a rainbow type effect. This is what would explain the halo effect.

Figure 1.7. Halo around the light source in this photograph is not even centered on the light source. If halos were due to refraction by perfectly aligned ice crystals, the halo would always be centered on the light source. This suggests that the lens is separate from the light source and was not quite in the correct position, in this case. The light source is not perfectly round or yellow as the real sun would be so it is clearly a simulator.

Thus, sun simulators have been in operation for at least 100s of years and clearly at a time when human beings did not have the technology to fly, much less to produce a sun simulator, which shows that aliens have been at it for a very long time. They are the ones hiding Planet X's effect on the Sun and the earth and they are clearly the ones that have filled humanity's knowledge base with illogical lies. In addition, they seem to have the ability to manipulate the human mind and to direct intelligent people away from the truth, for how else could they have stopped scientists for thousands of years from figuring out the truth?

So, at a time when people are engrossed with the idea of aliens, well, they are here, they have been here for a very long time. They are liars and they are malevolent, they clearly do not have humanity's best interests at heart. What historical record do we have regarding these aliens? Well, the only historical book that seems to have anything to say about them is the Bible, where we are told of the serpent that came into the Garden of Eden and got Adam and Eve to choose to disobey God and who as a result took over the planet. The Bible also tells us that the name of this serpent being, i.e. a reptilian, was Lucifer and that he is the prince of the power of the air, in other words, he has telepathic powers, which explains why he can put humanity's scientists to sleep and keep the truth away from them for thousands of years. The Bible also tells us that he can appear as an angel of light, i.e. he can appear in different forms. So, we seem to have an alien who is a reptilian shape shifter with telepathic powers who has taken over our planet thousands of years ago. The Bible also calls him the father of lies and tells us that he comes only to steal, kill and destroy. He most certainly has stolen the planet and the truth from us and the evidence that he is a liar is all around us, in the form of the illogical lies that passes as scientific research and the search for the truth about how the universe works.

In conclusion, aliens are clearly here and they are liars and malevolent, they have filled humanity's knowledge base with illogical lies and have been at it for thousands of years. These aliens are our enemies who seek only to destroy us.

References:

[1]   Albers, C. (2019). Article 785: Planet X is here but what is it exactly? (Book 12)

Chapter 2

1105. Earth's radius increased by 82% during the Flood cataclysm due to Planet X

Planet X seems to have first entered the Solar System several thousand years ago and its first approach to earth seems to have led to the Flood cataclysm, which caused the whole planet to flood with water and also caused it to expand, so that new lower elevation surface or basins appeared. The flood water then flowed into the newly formed basins, which thus turned into earth's oceans as this is what best fits the Bible account of the flood waters receding over a period of several months and also agrees with the evidence as viewed through Klauss Vogel's models of an expending earth.

Figure 2.1. Klauss Vogel's models of an expanding earth are based on the continents fitting perfectly on a much smaller globe and can explain why the southern continents are more widely dispersed than the northern continents and why the northern continents were expected to move more westward relative to the southern continents, i.e. the northern continents were closer to the north pole than the southern continents, land masses closer to the equator would move further apart.

Figure 2.2. National Geographic map of the Arctic Ocean: The ocean basin is the lower elevation region which seems to have expanded away from the mid ocean ridge. The higher elevations are the old continental shelves or the old surface of the planet.

The newly formed basin surface covered in water in the Arctic appears to be 50% of the total Arctic Ocean surface area.

Figure 2.3. Atlantic Ocean Floor: it almost all entirely made up of the ocean basin and thus newly formed surface due to the Flood cataclysm.

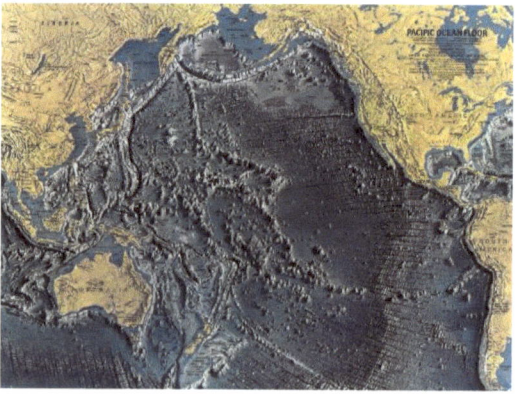

Figure 2.4. The Pacific Ocean, the largest ocean on the planet, is also almost entirely made of up of the ocean basin which formed at the Flood.

Earth's oceans cover 71% of the earth's surface, so it is likely that 70% of the earth's surface is newly formed surface. Thus, 30% of the earth's current surface was the surface the planet had before the Flood. We can use that to calculate the earth's radius before the flood, since we have that

$$A_{bf} = 0.3 A_c \tag{1}$$

where A_{bf} is the planet's surface area before the Flood and A_c is the planet's current surface area. Then since the area of a sphere is given by: $4\pi r^2$, we have that

$$4\pi r_{bf}^2 = 0.3(4\pi r_c^2) \quad \Rightarrow \quad r_{bf} = \sqrt{0.3}\, r_c = 0.55 r_c = 3490 \text{ km} \tag{2}$$

where r_{bf} is the planet's radius before the flood and r_c is the planet's current radius. Hence, the planet's original radius was 3490 km (2170 miles), or only 55% of what it is now, and thus the planet grew in terms of radius or diameter by 82%. This percentage is obtained from

$$\frac{r_c - r_{bf}}{r_{bf}} = \frac{0.45 r_c}{0.55 r_c} = 0.82 \tag{3}$$

In conclusion, the earth seems to have grown in terms of radius by 82% as a result of Planet X's initial effect on the planet, which resulted in the Flood Cataclysm, and in earth's current oceans forming. The planet's initial radius seems to have been only 55% of the current radius or 3490 km (2170 miles).

Chapter 3

1106. Moon's tidal pattern and the size of Earth's central core: 1745 km

The accepted size of earth's inner core is 1216 km, which would make it 19% of the total radius of the planet (6371 km). However, that would make earth's central core smaller than the moon. The moon cannot in any way be a natural earth satellite as central cores eject satellite cores which seem to be no larger than 100th the size of the central core, and the Moon has a radius of 1737 km, which makes it 27.3% the size of the earth, in terms of radius and thus too large to have been ejected by the earth's central core. The Moon seems to have been the first Planet X planetary central core to have come in and have forced itself onto the earth's core system and go through an energy absorption process that eventually allowed it to reach a state of energy equilibrium and thus to having a near circular orbit (see Article 1093: Space junk in orbit around the earth: Planet X cover-up and Article 1058: The Moon is a Planet X planetary central core and not a spaceship) [1, 2].

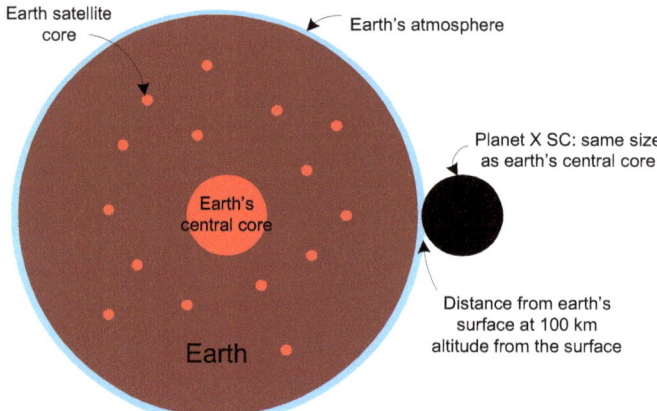

Figure 3.1. Planet X cores or Stellar Cores (SC) are able to very closely approach the surface of the earth when they first begin the energy extraction process from the earth's core system and thus follow extremely eccentric orbits. The planetary central cores seem to be about between 400 and 2400 km in radius, Mercury seems to be the largest known (see Article 1030: Mercury is not a planet: it is a part of the Planet X System) [3]. Mercury was adopted by the Sun, directly, but other smaller objects were observed orbiting the Sun inside Mercury's orbit in the 1800s (see Article 1028: Earth's new moons and size of Planet X objects observed in 1800s) [4]. These may have eventually become satellites to original Solar System Planets, of which there seem to only be 4: Venus, Earth, Mars and Pluto. All other planets and moons are Planet X cores. The Gas Giants are Planet X star central cores [3].

The Moon has a tidal effect on earth's oceans, which gives rise to recurring tides every 6 hours, everywhere, except in the most northern latitudes, which can only be explained if the moon's tidal pattern, on the earth's surface, has a central hollow and a tidal bulge ring. The tidal pattern may also be understood as a gravitational interference pattern, due to a gravitational connection between 2 cores.

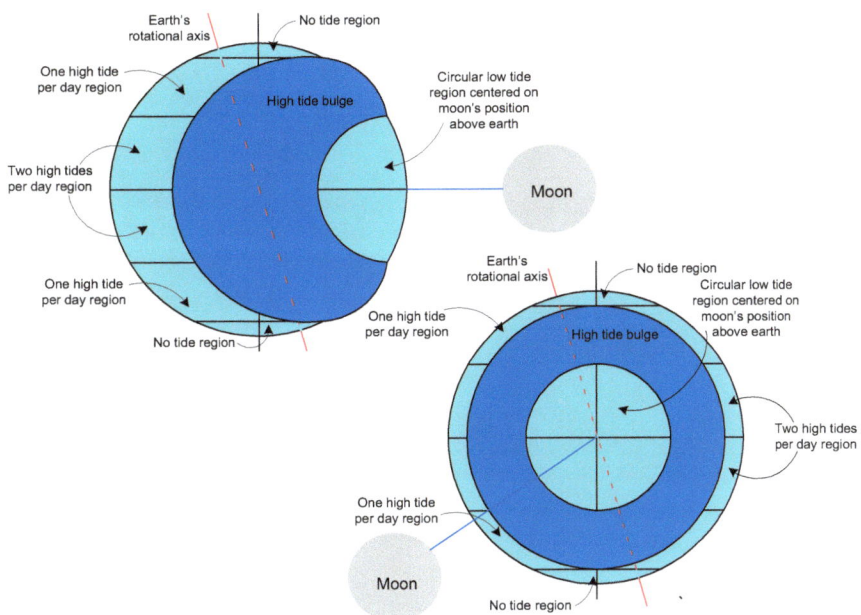

Figure 3.2. The moon's tidal pattern has to have a central hollow in order to create 6 hourly tides, at most lower latitudes and 12 hourly tides at the highest latitudes, i.e. close to the poles. The tidal bulge is also much larger than the moon itself, which has to be due to the fact that it is no longer energy depleted, like the Planet X central cores that are very closely approaching the surface of the earth, currently, and it therefore has a strong gravitational effect from its current position, at about 60 earth radii.

However, objects smaller than the earth, such as a tornado SC also produce gravitational interference patterns, on the earth's surface, which have an outer ring where earth matter is pulled upwards, and thus have central hollows, whilst other Planet X cores create tidal or gravitational interference patterns, which seem to have central bulges, which suggests that their position is inverted with respect to tornado and earth, i.e. they are larger than earth and are thus in its position, whilst the earth is in the tornado position because it is smaller.

Figure 3.3. The pattern on the surface of the earth has an outer bulge ring or a central hollow, the pattern on the surface of the tornado creating SC has a central bulge, thus the interference pattern associated to a gravitational connection between cores has central bulge on the smaller and a central hollow on the larger.

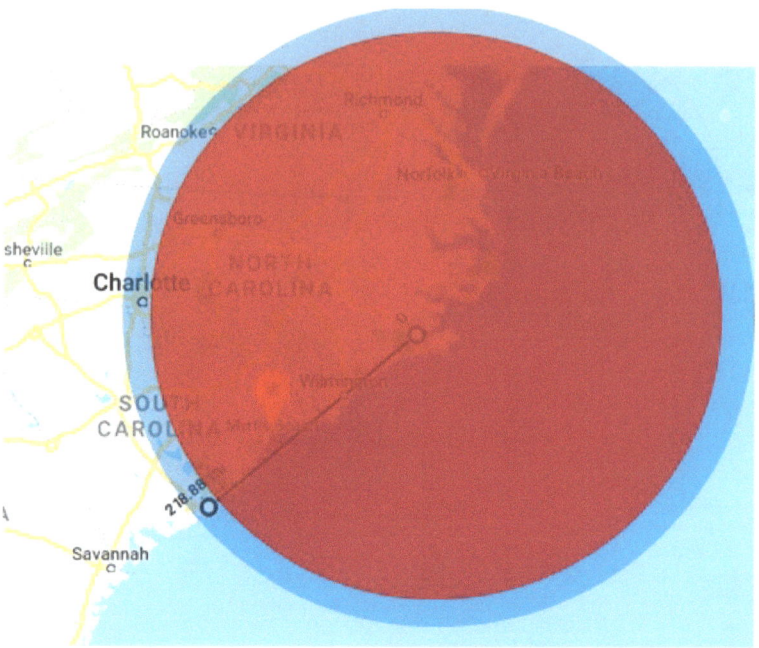

Figure 3.4. The gravitational interference pattern associated with Hurricane Florence and now with Hurricane Dorian as well, had a central bulge, which suggests that the earth's central core was the smaller of the two connecting cores or that the Planet X cores responsible were larger than earth's central cores.

Now, the moon creates a pattern with a central hollow on the earth's surface, which indicates that the earth's central core is the larger of the two cores. Thus, the earth's central core must be at least 1737 km in radius.

Then, since earth's original radius, before the Flood, was determined to have been 3490 km (see Article 1105: Earth's radius increased by 82% during the Flood cataclysm due to Planet X) [9], which would make earth's central core, if it was the same size as the moon, i.e. 1737 km in radius, 49.8%, i.e. very close to 50% of the size of the planet, this suggests that the normal size of a celestial object's central core is 50% of the size of that celestial object. With an earth core of exactly 50% the size of the planet, as it originally was before Planet X arrived in the Solar System, earth's central core should then have a radius of 1745 km.

Figure 3.5. Planet X central core inside the earth's lower atmosphere and thus in an extremely depleted state and still at the beginning of the energy process that will eventually see it settle in a close to circular orbit around the earth and creating its own tidal pattern on earth's oceans.

In conclusion, the moon produces a tidal pattern with a central hollow suggesting that the moon is smaller than earth's central core. Using this result and the earth's computed, before the flood, radius of 3490 km, it can be deduced that celestial objects, unaffected by Planet X energy depleted core systems seem to have central cores of 50% the size of the whole object, in terms of radius. This then suggests that the earth's central core has a radius of 1745 km. However, as we will find in a subsequent chapter this is not correct, the body of the earth is like an outer envelope for a planet's central core and thus in the case of the moon, which is now connected to the earth's central core gravitationally what counts is that the earth's overall radius is greater than its radius.

References:

[1]   Albers, C. (2019). Article 1093: Space junk in orbit around the earth: Planet X cover-up.
[2]   Albers, C. (2019). Article 1058: The Moon is a Planet X planetary central core and not a spaceship.
[3]   Albers, C. (2019). Article 1030: Mercury is not a planet: it is a part of the Planet X System.
[4]   Albers, C. (2019). Article 1028: Earth's new moons and size of Planet X objects observed in 1800s.

Chapter 4

1011. Planet X cataclysm: size and closeness of the objects responsible

Figure 1 below shows a photograph sent in by Sandra, of the sky over California, in which we can see the typical cloud formation associated with the presence of a huge Planet X object in the sky. The pattern is similar to the photographs sent in by Beth and Ryan, from two locations north of San Francisco (see Article 1005: Planet X in the sky and California earthquakes) [1]. Sandra's photograph was taken from south of San Francisco and is from July 11th and so from a few days before Beth's and Ryan's photographs. As in the other cases all we can still see is a flat surface in the sky covered in the clumpy cloud envelope, suggesting that this is indeed a huge object. In addition, there is evidence, in the photograph, of a connection point with cloud spout material, hanging down from it.

Figure 4.1. Photograph sent by Sandra from Ben Lomond, California near Santa Cruz, from July 11th 2019, 5 days before photographs of the sky, from north of San Francisco, also showing a flat surface covered in clumpy cloud, which is an indication that a huge Planet X object is in the sky. The cloud spout formation, in the midst of a clearing, is typical of the connection points, on these objects and is a sign that a gravitational connection has been made between the object and the earth's core, with the small spout being pulled down toward the center of the earth.

Figure 4.2. One of the first images, I saw that made me realize that these objects were in the sky, at very high altitudes, but certainly inside the atmosphere, or at least a part of the objects were penetrating the atmosphere down to the level where they could be observed and the artificial sky simulation system was not able to hide them. What seems to be a blue break in the clouds is actually the surface of a solid object and it is curved downwards, the cloud spout indicates this clearing to be a connection point. The object creates this type of cloud at the gravitational connection point and the spout is reminiscent of a tornado spout but much smaller. The cloud envelope cloud elsewhere is sparse and clumpy with lots of spaces through which the blue light emitting surface is visible. The 'Sun' seems to also be a reflection off the surface of the object.

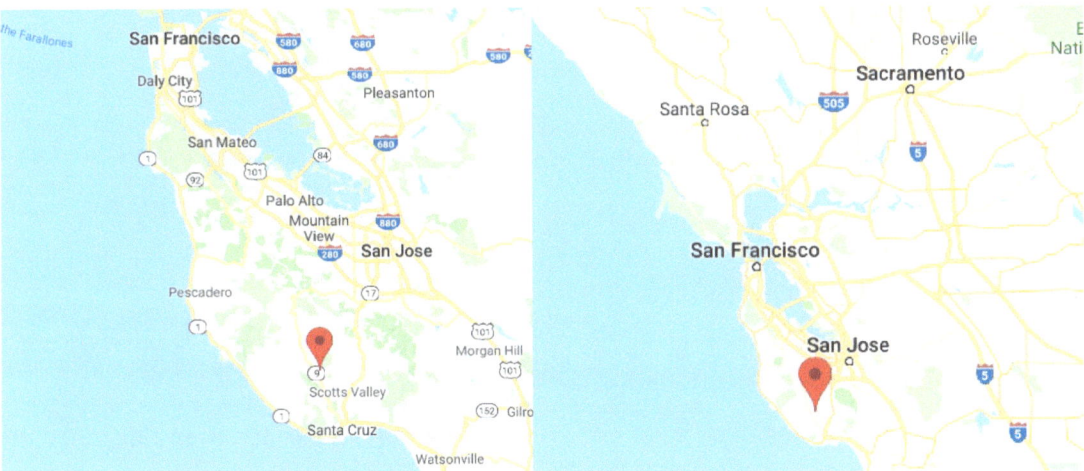

Figure 4.3. Sandra's photograph was taken from Ben Lomond and Beth's from close to Santa Rosa.

Figure 4.4. Right: Photograph by Sandra. **Left:** Photograph sent in by Beth. The cloud pattern is similar as we can see clumps, but the clumps seem to be a little different suggesting that this is not the same object. This suggests that there are more than one object, of this size, making connections with the earth's surface at this location close to San Francisco.

At the altitude that the objects' seem to be which appears to be higher in the case of the one in Beth's photograph on July 16th, it is likely that the surface we are seeing of the object is at least 200 km (120 miles) in width.

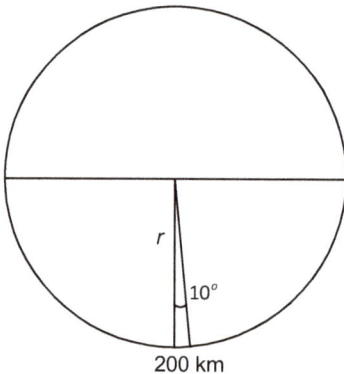

Figure 4.5. An object which seems to be a flat surface from the earth's surface over a distance of at least 200 km (120 miles) would mean that this distance is subtended by an angle no greater than 10°.

Thus, the object will have a radius:

$$r = \frac{s}{\theta}$$

Where s is the arc length (distance along a section of the circle's circumference) and θ is the angle in radians.

Hence,

$$r = \frac{200 \text{ km}}{10° \left(\frac{\pi \text{ rad}}{180°}\right)} = 1146 \text{ km} = 712 \text{ miles}$$

Therefore, this object will have an estimated medium radius of 712 miles, earth's central core is believed to have a radius of 760 miles, but since this is a minimum estimate the object could also be larger than the earth's central core.

Daylight in the earth's atmosphere is a consequence of the earth's atmosphere emitting light due to the energy flowing from the earth's core and only enough energy seems to flow through the atmosphere to allow light emission, when the atmosphere faces the sun, so the part of the atmosphere facing away from the sun does not emit light (see Article 888: The Sun is gone: Daylight comes from earth's core) [2]. This would mean that during the night there is not enough energy flowing through the earth's atmosphere to allow these objects to emit light, and so they will not be visible. They may remain connected at the same point though as earthquakes still occur during the night, or they may move to another connection point, after a few hours, thus giving rise to earthquakes and volcanic eruptions at different locations around the world.

Figure 4.6. Some more photographs from Sandra: On the left we see clearings on the surface and several lines in the clouds that indicate that the cloud layer may be following contours on the surface of the object. On the right we see a vertical patch which is most likely a narrow cloud spout connection.

Figure 4.7. Cloud patterns suggest contours on the surface of the object. Some of the lines suggest the presence of fissures and higher, than the surrounding surface, elevation features, i.e. mountain peaks.

These objects do not come down to as low an altitude as the smaller objects, which give rise to the weather but they must still be inside the lower atmosphere and thus below a 100 km (60 mile) altitude but this is still almost touching the surface of the earth in terms of the earth's radius.

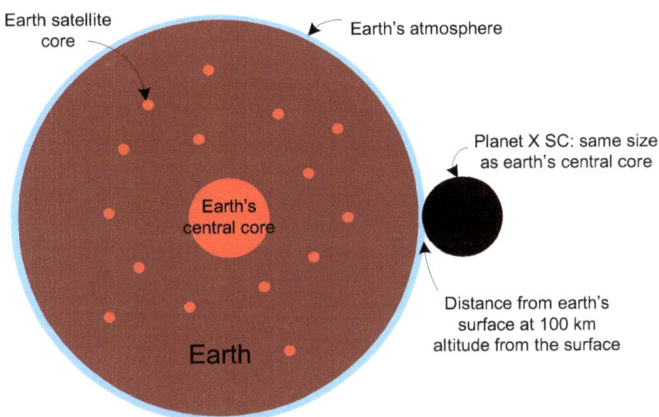

Figure 4.8. Illustration of how closely the Huge Planet X Stellar Cores, the energy depleted cores of the destroyed Planet X planets and stars, come to the earth's surface. An object of this size was most likely a central core in one of the Planet X planets. The objects come into the Solar System to absorb energy and thus attach themselves to the earth's core system, so that they become a part of the earth's core outside of the body of the earth. This object will most likely turn into an additional earth moon eventually. As they gain energy they seem to be more strongly repulsed by the earth's surface and move to higher orbits.

As they gain energy they seem to also form stronger gravitational connections and are thus more likely to generate ever stronger earthquakes and volcanic eruptions and as more and more of these objects come in and attach themselves to the earth, the earth's core becomes more and more unsettled and the type of reformation that is being seen on earth in California and Africa where rifts are opening overnight suggest that this cataclysmic event is now extremely close.

In conclusion, the observational evidence suggesting that there are huge Planet X Stellar Cores in the sky and very close to the earth's surface is irrefutable. The increased number and size of these objects points to a Planet X cataclysm being imminent.

References:

[1]   Albers, C. (2019). Article 1005: Planet X in the sky and California earthquakes (Book 13).

[2]   Albers, C. (2019). Article 888: The Sun is gone: Daylight comes from earth's core.

Chapter 5

1015. In search of Vulcan: Planet X observed traversing the Sun since 1818

The existence of planet Vulcan, a planet, which was supposed to orbit the Sun, inside the orbit of Mercury, was proposed by the French mathematician Urbain Le Verrier, in 1859. He proposed the existence of the planet because of an inconsistency in Mercury's orbit, where its perihelion position advances by 0.02° per year, a tiny amount, which can be nearly be accounted for by the influence of the other known planets, in the Solar System, leaving a discrepancy of 0.004° per year, which could be explained by an additional intra-Mercurial planet. Le Verrier had made a similar proposal regarding Uranus' orbit, which led to the discovery of the planet Neptune, an indication that the proposal was valid.

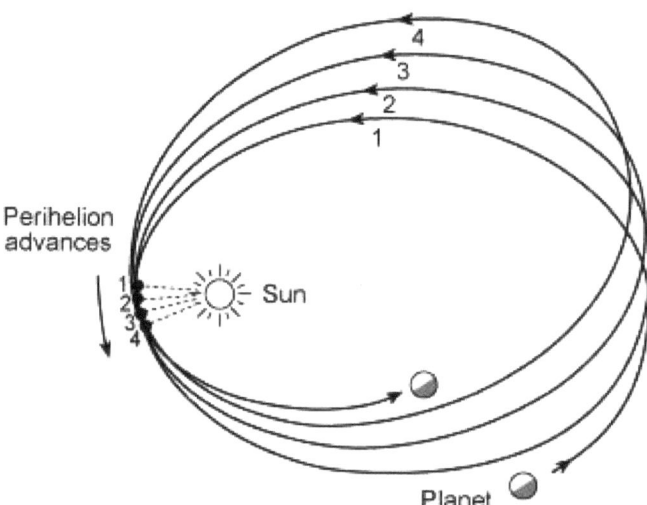

Figure 5.1. Mercury's orbit is quite eccentric and as a result the planet never retraces its orbit exactly, an effect which is called perihelion advance, the perihelion position of the orbit changes with every orbit.

On March 26 1859, a French physician and amateur astronomer Edmond Modeste Lescarbault observed through his 3.75 inch refracting telescope a black dot, on the Sun, which he took to be a sunspot at first, but then realized that it was moving and thus thought he was observing the transit of a previously undiscovered body, as Mercury was not in a position, where a transit could be observed from earth. He wrote to Le Verrier, who then visited and interviewed him, and was satisfied that he had observed the planet Vulcan. However, another astronomer, Emmanuel Liais, later discounted that such an observation could have been made as he claimed to have been observing the Sun at the same time as Lescarbault and had not seen the transit.

However, various other reports started emerging of objects being seen traversing the Sun. Capel Lofft reported seeing a body traversing the Sun on 6 January 1818. Gruithuisen reported seeing 2 small spots on the Sun, which were round and black and one was larger than the other, on 26 June 1819. Pastorff observed a transit on 23 October 1822, on 24 and 25 July 1823, 6 times in 1834, once on 18 October 1836 and on 1 November 1836; and on 16 February 1837, he saw 2 objects, the larger was 3 arc seconds across and the smaller 1.25 arc seconds across. These observations would suggest that there was not just one small planet inside the orbit of Mercury, and not just two either, but several, as it is impossible to observe the same object traversing the sun two days in the row, or even two weeks apart.

Then, on 29 January 1860, F.A. Russel and 3 other people saw a transit of an object. The fact that 3 people observed the same thing makes this an undeniable observation, but in addition, another astronomer, Richard Covington, also claimed years later to have observed a well-defined black spot traverse the Sun, in 1860, which gives this observation even more validity. Then, on 22 March 1862, another amateur astronomer, Mr. Lummis of Manchester, England, saw a transit and so did a colleague of his. Based on these two men's observations, two French astronomers, Benjamin Valz and Rudolph Radau computed the orbital period of the object and obtained extremely low values. Valz calculated it to be 17 days and 13 hours, Radau calculated it to be 19 days and 22 hours, which would place the object at between 0.13 au and 0.14 au, i.e. at 13 million miles from the Sun, which is twice the distance from the center of the Sun to the edge of the Sun's outer corona.

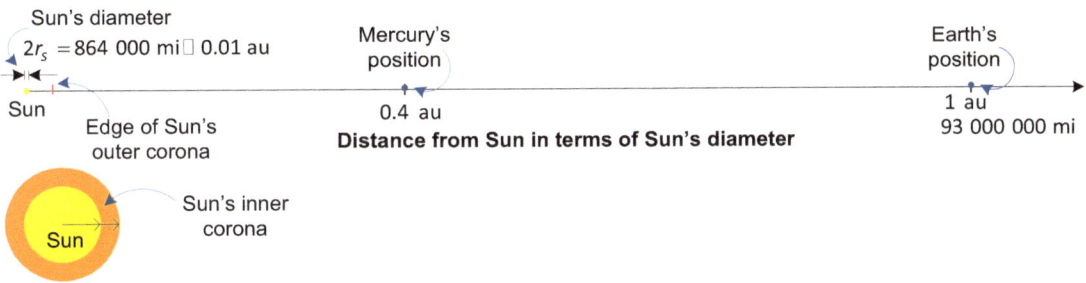

Figure 5.2. Illustration of the distance between the Sun, the edge of the Sun's outer corona, Mercury and Earth, in terms of the Sun's size. The Sun's size is tiny in comparisson with the distance between it and the Earth. The distance between the Earth and the Sun is about 100 times larger than the Sun's diameter. The distance between the Mercury and the Sun is about 40 % of the distance between the Earth and the Sun, or 0.4 au, where 1 au is the distance between the Earth and the Sun, in astronomical units. The Sun's outer corona goes out to a distance which is 12 times the radius of the Sun, or 5.2 million miles, which is equivalent to 0.06 au and thus Mercury is well outside the Sun's corona. The object observed traversing the Sun would be around where the blue arrow pointing to the edge of the Sun's outer corona crosses the axis. It would be much much closer to the Sun then Mercury.

On 8 May 1865, another French astronomer, Aristiche Coumbary, observed an unexpected transit from Istanbul, Turkey. And, on July 29 1878, 2 experienced astronomers, Prof James Craig Watson, the director of the Ann Arbor Observatory, in Michigan, US, and Lewis Swift, an amateur from Rochester, New York, US, both reported seeing a Vulcan type planet close to the Sun during a total solar eclipse. These two men were excellent observers, Watson had discovered 20 asteroids, while Swift had discovered several comets, which were therefore named after him, and both described the planet as red and close to superior conjunction; in other words, the planet was close to being a fully illuminated disk, with only one small portion darkened, on the right hand side. Skeptics claimed that the two men had mistaken known stars for planets. But it would be illogical to expect two men to observe an object that is similarly red and approaching superior conjunction, and be mistaken, especially since they were such experienced and excellent observers. Skeptics are still around today and are usually similarly disadvantaged in terms of ability to think logically.

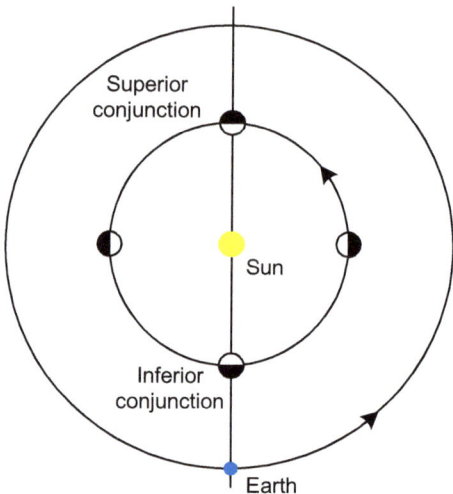

Figure 5.3. A planet inside the earth's orbit, called an inferior planet, is in superior conjunction when it is aligned with the earth and the Sun, but is on the opposite side of the Sun, and will thus be observed as a fully illuminated disk, from earth. A planet close to superior conjunction is a little to the right of the sun and thus a small part of the disk will be dark.

Then in 1915, Einstein proposed his theory of General Relativity, a theory of gravity which identifies gravity as curvature of space, which supposedly causes light waves to follow curved paths. This theory was used to explain the last tiny discrepancy in Mercury's perihelion advance. However, Einstein's theory is obviously incorrect as the very idea that space is curved is illogical. Space and frames of reference are mathematical constructs used to determine the position of matter, or of an event involving energy flow, or energy transformation, in the universe. They are not real, only matter and energy are real substances, in the universe. So, it should not be surprising that Einstein's theory of gravity can easily be shown to be incorrect with one simple example. According to Einstein's theory gravity curves space and all matter follows a path through this curved space, as if it is a straight line, i.e. the closest distance between 2 points in a curved space is like the closest distance between two points on a spherical surface. You cannot move through the surface, so you are forced to follow a curved path on the surface.

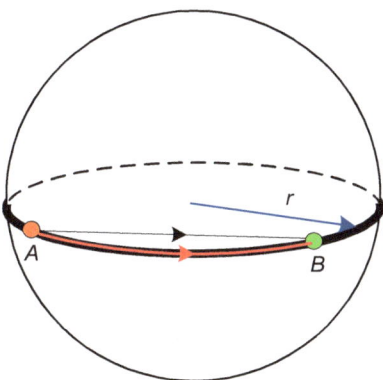

Figure 5.4. Curvature of space in Einstein's theory is like being forced to move on a curved surface, the closest position between two points is a curved path because it is not possible to move through the solid surface.

So, according to Einstein's theory space is curved around a massive object, and that is why spacecraft follow a curved path around it, the curved path is the closest distance between two points. But then a rope stretched between two spacecraft, one following close behind the other, in the same orbit, should be curved as well, as the curved path is supposed to be the closest possible distance between the two spacecraft, but it clearly is not. Common sense tells us that we can stretch a rope between the two spaceships and it will follow a straight line, not the curvature of the orbit, which shows that space is not curved; space is flat everywhere, thus completely falsifying Einstein's theory of gravity.

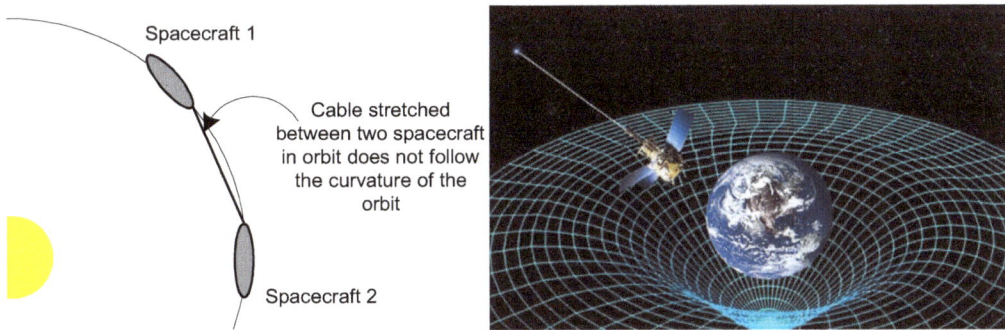

Figure 5.5. Left: A cable stretched between two spacecraft would be in a straight line, and would not follow the curvature of the orbit, which shows that gravity cannot be due to the curvature of space. **Right:** Two dimensional illustration of a three dimensional curved space as used in General Relativity.

However, even if Einstein's theory had been correct, the explanation of the discrepancy in Mercury's orbit, cannot, in any way, do away with the huge number of credible observations of objects traversing the Sun. Theory is falsifiable but observation is not. The very idea that it can, is illogical. Theory is supposed to help us understand the universe, observation gives us information about the universe we live in, observation cannot be done away with and theory should not be able to even be done without it as it has to strictly adhere to it in order to be valid and credible. But this is exactly what has been done, in this case, observation was suppressed and ignored in a favor of a theory, which is completely illogical and contrary to observation. But, the bolstering of wrong and illogical theories in order to cover up the truth seems to be something that is done over and over again, in scientific circles, which has caused humanity's knowledge base to be filled with lies and illogical inconsistencies. See Article 1040: Aliens have filled humanity's knowledge base with lies [1], where I show through a simple calculation that the time honored explanation about the Sun's role in Spring tides is impossible and thus a complete lie.

It seems that a special effort has therefore been made, on this planet, to hide the truth about gravity and Planet X, which was what all those Vulcan observations seem to have been about. And, much more recent Planet X observations actually reveal that gravity is a creative force and that there is continuous matter creation all over the universe, as a result, something which the astronomer Halton Arp showed many years ago, and for his efforts had his papers blocked from publication and was refused time at the telescopes, which clearly shows that an effort has been made to hide the truth (see Article 126: White Holes instead of Black Holes at the Center of Galaxies, Article 987: Trovants: rocks that grow: core matter, water and human consciousness and Article 988: Trovants, Planet X and how planets continuously grow in size) [2, 3, 4].

In the case of the Vulcan observations, we have clear observational evidence, from 1818, of objects traversing the Sun and being very close to the Sun. The observations reveal that it could not just have been one, there was clearly more than one. What were these objects? They were Planet X Stellar Cores, the name I gave to the system of cores, from destroyed planets and stars (see Article 785: Planet X is here but what is it exactly?) [5] that seem to have started affecting our planet at the time of the Flood, and that were responsible for the cataclysm that then occurred, and that are responsible for the cataclysm, which is now unfolding, on earth, and that will imminently reach a pinnacle of severity, which is likely to be just as cataclysmic as the Flood event but without completely flooding the earth. Instead, there will be extensive surface reformation, lava flows, volcanic eruptions and cataclysmic earthquakes and thus tsunamis accompanied by increasingly severe weather events.

In conclusion, there is clear observational evidence of objects inside of Mercury's orbit traversing the Sun, and not just of one of them, but at least several. So, what started as the search for one small planet, named Vulcan, ended up giving extensive and credible observational evidence for the existence of Planet X Stellar Cores, the energy depleted cores of the destroyed Planet X planets and stars, near the Sun. In addition, it seems clear that extraordinary efforts have been made to hide the truth regarding the existence of Planet X and the truth regarding the gravitational interaction, which is in fact a creative force, leading to matter being continuously created all over the universe.

References:

[1] Albers, C. (2019). Article 1040: Aliens have filled humanity's knowledge base with lies.
[2] Albers, C. (2019). Article 126: White Holes instead of Black Holes at the Center of Galaxies.
[3] Albers, C. (2019). Article 987: Trovants: rocks that grow: core matter, water and human consciousness (Book 13).
[4] Albers, C. (2019). Article 988: Trovants, Planet X and how planets continuously grow in size (Book 13).
[5] Albers, C. (2019). Article 785: Planet X is here but what is it exactly? (Book 12)

Chapter 6

1028. Earth's new moons and size of Planet X objects observed in 1800s

In Article 1015: In search of Vulcan: Planet X observed traversing the Sun since 1818 [1], I detailed the evidence which clearly shows that several objects were observed traversing the Sun, inside Mercury's orbit, starting from 1818. One of these observations occurred on 22 March 1862 by an amateur astronomer, Mr. Lummis of Manchester, England, and a colleague of his, and based on this observation, two French astronomers calculated the orbit of the object, and thus determined that the object was between 0.13 and 0.14 au from the Sun, at the time.

Figure 6.1. The object would be at about the point where the arrow, indicating the edge of the Sun's outer corona, crosses the axis. This is a very close distance to the Sun. 1 au is 150 million kilometers or 93 million miles.

Then, on 16 February 1837, another astronomer observed 2 objects traversing the sun and determined that one of the objects had an angular width of 3 arc seconds and the other 1.25 arc seconds. So, using the distance from the Sun calculated by the two French astronomers, and thus assuming that these objects were at 0.14 au from the Sun, we can estimate the diameter of these objects using the relationship between arc length and angular width:

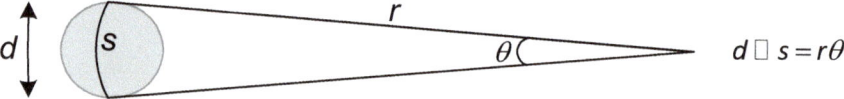

Figure 6.2. The object would be at a distance $r = (1 - 0.14)$ au $= 0.86$ au from the earth, so its diameter can be determined from: $d = r\vartheta$, where ϑ is in radians.

Hence, the two objects would have diameters:

$$d_{1.25} = (0.86 \text{ au})\left(\frac{150 \times 10^6 \text{ km}}{1 \text{ au}}\right)(1.25 \text{ arc sec})\left(\frac{1°}{3600 \text{ arc sec}}\right)\left(\frac{\pi \text{ rad}}{180°}\right) = 782 \text{ km}$$

$$d_3 = \frac{3}{1.25}d_{1.25} = 1876 \text{ km}$$

Thus the larger of these two objects would have a radius of 938 km, a size which is very close to the 1000 km radius estimated to be the size of the huge Planet X Stellar Cores (SCs) which are coming into the earth's atmosphere and creating flat surfaces in the sky (see Article 1011: Planet X cataclysm: size and closeness of the objects responsible) [2].

Figure 6.3. Flat solid surface in the sky above the earth's surface created by the presence of a huge Planet X SC inside the earth's atmosphere: The cloud envelope is sparse and clumpy and the Sun appears to be a reflection off the surface of the object.

Figure 6.4. The cloud pattern above is similar to this one over Hawaii from June 20th 2019. The Sun is a reflection off the surface and the clumpy clouds indicate that we are looking at the solid blue light emitting surface of a large Planet X SC (see Article 992: Huge Planet X object in the sky over Hawaii: causes dizziness and nausea) [4].

Figure 6.5. The Moon: a Planet X central SC, which came into the earth, at the time of the Flood cataclysm and which all the Planet X SCs will most likely look like. The additional huge SCs which are now coming into the earth's atmosphere will thus look like moons and are most likely orbiting the earth as moons but with very elliptical orbits which brings them deep inside the earth's atmosphere at times.

In conclusion, a Planet X object observed traversing the Sun in the 1800s, has an estimated radius of 938 km (582 miles), a size which is close to the estimated size of the huge Planet X objects observed in the earth's skies, which create the appearance of solid surfaces at altitudes of less than 100 km. These objects will most likely look like moons and be in orbit and in very elliptical orbits around the earth so that the earth will now have several new moons.

References:

[1]   Albers, C. (2019). Article 1015: In search of Vulcan: Planet X observed traversing the Sun since 1818.

[2]   Albers, C. (2019). Article 1011: Planet X cataclysm: size and closeness of the objects responsible.

Chapter 7

1030. Mercury is not a planet: it is a part of the Planet X System

As I have shown in Article 1029: Planet X as new moons orbiting the earth: the irrefutable evidence [1], the earth has new moons and these moons look a lot like the earth's original moon, in other words, they have a dark greyish cratered surface. Planet X within the Solar System clearly leads to the formation of volcanoes on the surface of Solar System planets, which is why both Mars and Earth have volcanoes, but the moon does not have volcanoes because it is not a planet, and thus does not have a core system inside it. Instead, the moon is a core, most likely a central core for a Planet X planet, which broke up thousands of years ago. But Mercury looks a lot like the moon, which suggests that it too is a core or a Planet X central Stellar Core, the name I came up with for these objects.

Figure 7.1. Left: Mercury. **Right**: Moon. Both objects look very much alike and neither have volcanoes on their surface, which we would expect, if they were whole celestial objects, as whole celestial objects would have core systems inside them, which create different materials such as water, soil and magma. Both seem to therefore be Planet X central Stellar Cores, the energy depleted central cores of Planet X planets, which come into the solar system to absorb energy and become a part of either the Sun's, or one of its planets', core system.

Now, Mercury seems to have first been observed in 265 BC by Timocharis, but its orbit was only studied in 1639, by Zupus, and it was not until the 1960s that its rotational period was determined because due to its closeness to the sun, it was too difficult to discern surface features. Then, since several objects were observed traversing the Sun, which would be either as close to the Sun as Mercury, or closer, since 1818 (see Article 1015: In search of Vulcan: Planet X observed traversing the Sun since 1818) [2], it is possible that for a very long time different objects were observed and taken to be one object, i.e. Mercury. Two of the objects observed traversing the Sun since the 1818, by which time it was understood that they could not have been Mercury, were estimated to be smaller than Mercury (391 km and 978 km radii) but it is possible that there are some which are as large as, or larger than, Mercury (see Article 1028: Earth's new moons and size of Planet X objects observed in 1800s) [3].

In fact, even though the Moon and Mercury orbit different Solar System objects their orbits are quite similar as both have orbits that are far from circular and also inclined with respect to the ecliptic suggesting that they are not original Solar System objects but have been around long enough to have settled in close to stable orbits. Since Mercury has a radius of 2440 km and the moon has a radius of 1737 km, the reason why Mercury ended up orbiting the Sun rather than the earth or another of the Sun's planets is likely to be because Mercury is not compatible with the earth's core system, in other words, it is larger than the earth's central core, but the moon did, which then suggests that the earth's central core is at least as large as the moon and thus has a larger radius than the current accepted value of 1220 km.

Figure 7.2. Image of Jupiter, in ultraviolet light, after Shoemaker Levy 9 impacted it, in July of 1994, showing the large blemishes that were left on Jupiter, which suggests a large electrical interaction and that there was a solid surface right underneath the top of the gaseous layer. The other evidence which also suggests a mainly solid object is that all Gas Giants obey the same relationship between their radii and orbital radius ($r \propto R^{-3/2}$), which would only be possible if they all have the same density and are very similar, i.e. they have to be solid cores with a very small outer gaseous layer (see Article 621: Gas Giant vindication and why Dr Eugene Shoemaker was really killed) [4].

Now, Jupiter, as well as, all the Gas Giant planets, seem to also be Planet X SCs but they would be star central cores, rather than planetary central cores, and the fact that they are so much larger than Mercury suggests that planetary central cores and star central cores are fundamentally different. This is more likely as a result of the repulsive gravitational force, which keeps cores or any celestial object from colliding, being much stronger, in larger central cores. So, star central cores end up much further from each other, whilst planets and thus planetary cores are able to approach other celestial objects much more closely, and thus end up in stable orbits, which are much closer to the host star than star cores.

In conclusion, the discovery that Planet X objects smaller than Mercury were observed traversing the Sun since 1818 has led to the understanding that Mercury is also a Planet X Stellar Core, and not an original Solar System planet like earth and Mars. Mercury seems to thus be a central core of one of the destroyed Planet X planets. Also, Planet X star cores, of which the known Gas Giants, in the Solar System, seem to be examples, end up orbiting the Sun much further than the Planet X planetary cores because the repulsive part of the gravitational interaction seems to be much stronger in the larger cores.

References:

[1] Albers, C. (2019). Article 1029: Planet X as new moons orbiting the earth: the irrefutable evidence (Book 15).

[2] Albers, C. (2019). Article 1015: In search of Vulcan: Planet X observed traversing the Sun since 1818.

[3] Albers, C. (2019). Article 1028: Earth's new moons and size of Planet X objects observed in 1800s.

[4] Albers, C. (2019). Article 621: Gas Giant vindication and why Dr Eugene Shoemaker was really killed.

Chapter 8

1058. The Moon is a Planet X planetary central core and not a spaceship

Emmanuel Velikovsky collected extensive research, which clearly showed that the moon had not always been in our skies. He found that the Greek philosopher, Aristotle, for example, mentions that Arcadia, in Greece, at a time before it was inhabited by the Hellenes, was inhabited by a people called Proselenes, when there was no moon in the sky. Also, Plutarch, another Greek philosopher, wrote about the Arcadians as being pre-Lunar people [1]. So it is clear that the moon has not been around for more than a few thousand years and it was thus not around before the flood. But that does not mean that it is a spaceship because the moon has tidal effects on the earth's oceans, which requires a gravitational interaction between two central cores. If the moon had been a spaceship it would not be able to make a gravitational connection with the earth's central core and thus create tides.

Figure 8.1. The rounded shape of Birch Bay suggests that it originally formed as a result of Planet X making a connection at this location, which led to the formation of a sinkhole. The shape of the ground under the water appears to be in the form of 2 other circular outlines indicating that another 2 smaller sinkholes appear to have formed within the larger outline of the Bay, which suggests that smaller objects have made similar connections at this location (see Article 1000: Birch Bay tides connected to Seattle earthquakes and due to Planet X) [2].

Birch Bay is also the location where an extreme low tide in June of 2019 made me aware of this location being a point where at least one Planet X central core was coming on a regular basis to make a gravitational connection, with the earth's central core, which then resulted in the abnormally low tide.

Figure 8.2. A much lower tide than normal in Birch Bay in June of 2019 led to the discovery that this is a Planet X planetary central core connection point with the earth's central core.

The powers that be had tried to cover-up what had been going on regarding these exceptionally low or high tides, for thousands of years, by explaining these as being due to the moon and the Sun being aligned, but a simple calculation shows that the sun's tidal force is only 0.25% that of the moon's and cannot thus in any way cause a tide which is 30% higher or lower than normal (see Article 936: Birch Bay extreme low tide not due to a spring tide) [3]. But in fact all tides must be due to the same type of gravitational connections because gravity must work in the same way in all cases, so if it is Planet X central cores making gravitational connections with the earth's central core, which are causing the more local but much more severe tidal effects, then even those tides that are caused by the moon must be produced through the same process, which thus shows what the moon is. The moon is a Planet X central core.

In addition, the moon appears to be about the maximum size of the Planet X central cores that are compatible with the earth's core, since Mercury, which is clearly a Planet X planetary central core as well but a bit larger than the moon ended up orbiting the Sun rather than one of its planets (see Article 1030: Mercury is not a planet: it is a part of the Planet X System) [4]. Since planets seem to be ejected from stars and must therefore be star satellite cores and the earth is about 100 times smaller than the sun, it is likely that satellite cores are at least 100 times smaller than the central core, in a system, and thus all of the earth's satellite cores would not be any larger than 100th of the size of earth's central core. Since earth's central core is most likely the same size as the moon, which has a radius of 1737 km, earth's satellite cores, would be no larger than 17 km or 11 miles in radius, and thus no larger than 22 miles in diameter. But most will be smaller than that. This means that there is no process for planets having natural moons at least not with a size as we find moons to have in the Solar System. This therefore indicates that all the moons across the Solar System are in fact Planet X planetary central cores, which were once inside the Planet X planets, before they were destroyed.

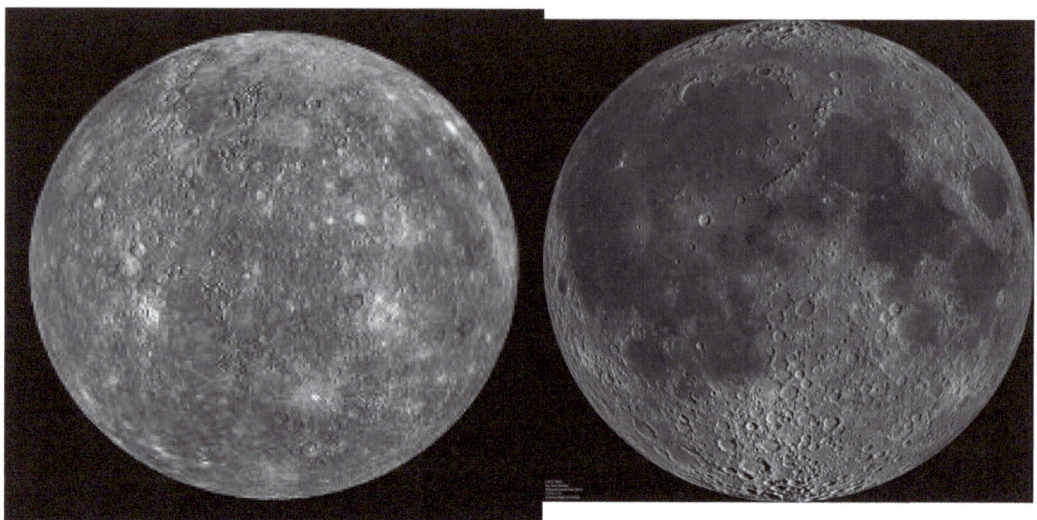

Figure 8.3. Left: Mercury. **Right**: Moon. Both objects look very much alike and neither have volcanoes on their surface, which we would expect, if they were whole celestial objects, as whole celestial objects would have core systems inside them, which create different materials such as water, soil and magma. Both seem to therefore be Planet X planetary central Stellar Cores, the energy depleted central cores of Planet X planets, which come into the solar system to absorb energy and become a part of either the Sun's, or one of its planets', core system.

This also explains why both the Moon and Mercury have eccentric orbits with a high inclination, with respect to the ecliptic. They have most likely been around for several thousand years but have not yet managed to settle in a perfect circular orbit, perfectly aligned with the ecliptic plane, and also suggest that their orbits were once much more eccentric than they are now. And in fact, in order for the Planet X central cores to regularly come inside the earth's atmosphere and make gravitational connections they would have to have an extremely eccentric orbit and most likely even more inclined with respect to the ecliptic plane than the moon's or Mercury's orbit.

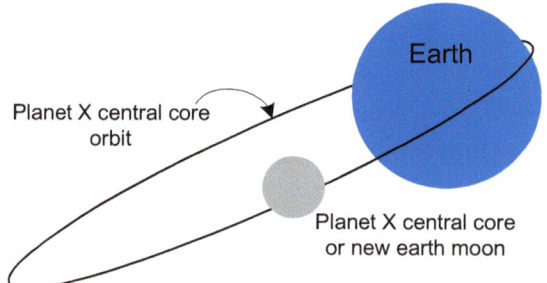

Figure 8.4. The type of orbit that earth's new moons or Planet X planetary central cores are likely to follow, at least initially: extremely eccentric and deeply inclined with respect to the ecliptic. They may go out as far as the Moon's orbit is but the perihelion of their orbit brings them inside the earth's lower atmosphere where they are able to remain stationary for a time, rotating with the earth, whilst making a gravitational connection at a certain point on the earth's surface. They are also likely to come back to the very same point over and over again.

When they come back to make a connection they must be facing in exactly the same direction as last time, which also explains why they would tend to become tidally locked as the moon is, i.e. the same side always faces the planet surface.

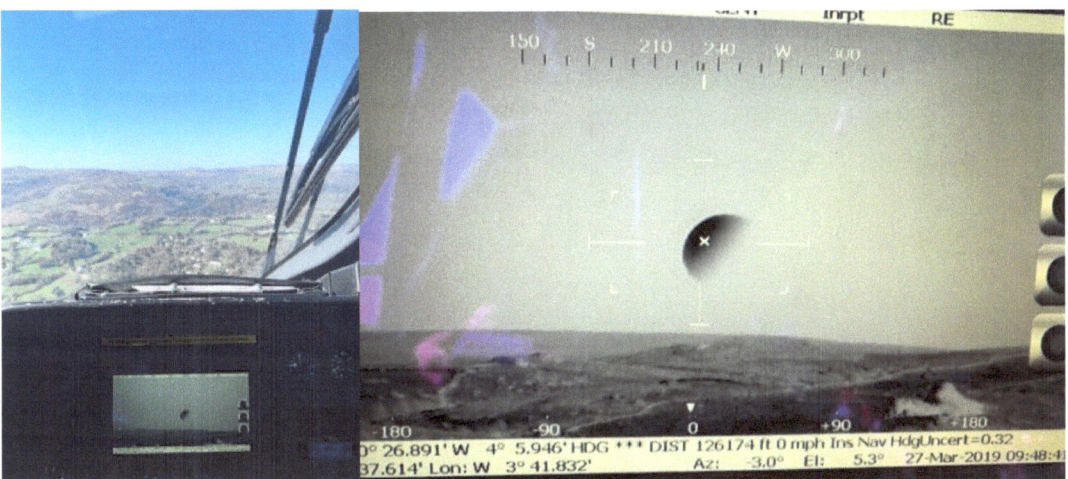

Figure 8.5. One of earth's new moons is here visible through a helicopter's thermal or infrared light camera over Dartmoor, England, on 27 March 2019 (see Article 919: The Earth has more than one Moon: confirmed) [5].

When placed next to the earth's original Moon, it is obvious that it is not the same Moon, but yet it looks like a moon as it has a cratered surface like earth's original Moon.

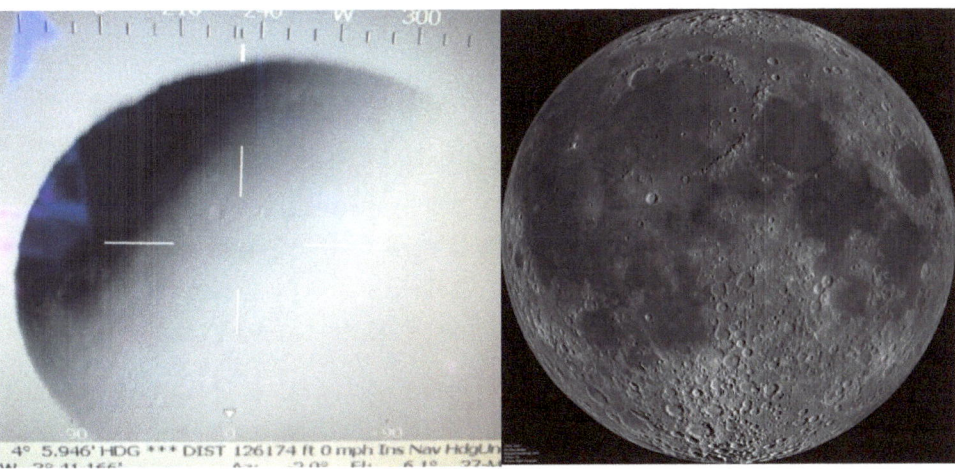

Figure 8.6. When the Dartmoor moon image is placed next to the image of earth's Moon, earth's original satellite, it becomes obvious that this moon is far from being as spherical as the Moon. This new moon has large indentations in its outline suggesting that it is much smaller and much closer than earth's original Moon.

But this is not the only one, earth seems to have several of these objects in orbit, which come into the earth's atmosphere and make gravitational connections with several undesirable consequences such as volcanic eruptions, earthquakes, tidal effects, huge waves, severe storms, flash floods, fires, power grid failure, mass animal deaths. In addition, the debris which they bring with them seems to also bring diseases and insect plagues with them (see Article 1053: Tick population and associated Lyme disease on the rise due to Planet X) [6].

Figure 8.7. Blue light emitting solid surface, in the sky, which can only be produced by a huge solid object inside the earth's atmosphere, or at least, by part of its surface being inside the earth's atmosphere: The objects clearly create the cloud, which appears to sparsely cover the surface of the object and that, in this case, is emitting pink light. They create a different type of cloud at the connection points, which I have named cloud spout material. The clearing observed here is one of these connection points. It may simply look like a very large crater on the surface of the object once it is outside the earth's atmosphere (see Article 1029: Planet X as new moons orbiting the earth: the irrefutable evidence) [7].

Hence, the fact that the earth has new moons that look like earth's new moon and that the moon is so similar to Mercury, which is clearly also a Planet X planetary central core, and also the fact that the Moon and Mercury have such similar orbits can only lead to one possible logical conclusion, earth's original moon is a Planet X planetary central core and since it must have come in, at about the time of the Flood, it is most likely the object associated with the cataclysmic event that occurred then.

In conclusion, the earth not only has new moons, which are Planet X planetary central cores, earth's original Moon is clearly a Planet X planetary central core as well, and cannot in any way be a spaceship as a spaceship would not be able to make a gravitational connection with the earth's central core and create tidal effects, it takes a natural core, an object that was ejected by a star's core system.

References:

[1] Velilovsky, I. (1940s). The earth without the Moon: https://www.varchive.org/itb/sansmoon.htm

[2] Albers, C. (2019). Article 1000: Birch Bay tides connected to Seattle earthquakes and due to Planet X.

[3] Albers, C. (2019). Article 936: Birch Bay extreme low tide not due to a spring tide.

[4] Albers, C. (2019). Article 1030: Mercury is not a planet: it is a part of the Planet X System.

[5] Albers, C. (2019). Article 919: The Earth has more than one Moon: confirmed.

[6] Albers, C. (2019). Article 1053: Tick population and associated Lyme disease on the rise due to Planet X.

[7] Albers, C. (2019). Article 1029: Planet X as new moons orbiting the earth: the irrefutable evidence.

Chapter 9

1093. Space junk in orbit around the earth: Planet X cover-up

The earth is supposedly surrounded by a manmade debris field. This space junk or space debris is supposed to be due to satellites, which have malfunctioned and may have then collided with pieces of debris or other satellites, which would break up into pieces and thus turn into debris themselves. However, malfunctioning satellites will eventually re-enter the atmosphere as their orbit will decay and thus collide with the earth's surface eventually, and space is a very large place, so to suggest that the earth is literally surrounded in a dense cloud of debris due to artificial satellites being launched into orbit is illogical. Even at a rate of 100 satellites per year, only 5400 satellites would have been launched since 1961, which is hardly enough to cause a cloud of debris around the earth especially since the oldest lower altitude satellites would have re-entered the atmosphere.

Figure 9.1. Image illustrating the amount of manmade space debris in orbit around the earth, which is illogical. It does not seem possible that human beings have placed this many satellites in orbit since the 1960s. It is more likely that the debris field is due to Planet X and thus natural.

The use of an illogical story-line is usually a sign that it is an attempt to cover-up the existence of Planet X. The 'powers that be' are so terrified that the earth's population will find out the truth about Planet X that they have spent thousands of years filling humanity's knowledge base with lies (see Article 1040: Aliens have filled humanity's knowledge base with lies) [1]. And since Planet X comes in as systems with a central core and satellite cores, which may number in the 100s or even thousands and since the central core, after a period of time spent absorbing energy, in an eccentric orbit, eventually, goes into a circular orbit, we would expect nothing less from the satellite cores, which means that there should now be a huge number of these cores in close to circular orbits around the earth.

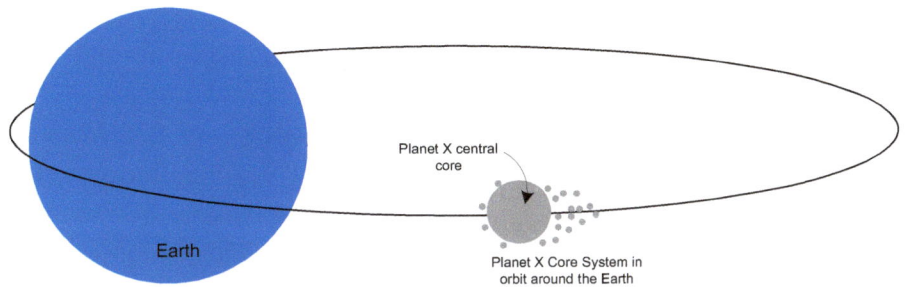

Figure 9.2. Planet X System in orbit around the earth. The system is made up of one central core surrounded by 1000s of satellite cores, which are likely to be 100th the size of the central core or smaller.

The fact that satellite cores as well as central cores end up in near circular orbit can be seen from Mars, since it has two very small moons in orbit, Phobos has an average radius of 11 km and Deimos has an average radius of 6.3 km. The orbital period of both moons is exactly the same as their rotational period, just like earth's original Moon, which means that they always present the exact same side to the planet, something, which is referred to as being tidally locked. But it is because they are Planet X cores, and gravitationally connected to the planet's central core. Planet X cores make gravitational connections at connection points, on their surfaces, and we would thus expect such a point on the surface of these objects to always point toward the surface of the planet.

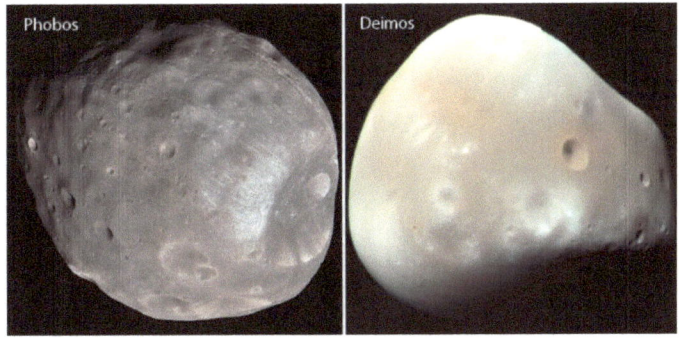

Figure 9.3. Mars' moons were discovered in 1877 and they are very small, Phobos the closest, which only orbits about 6000 km from the planet's surface has an average radius of 11 km and Deimos, which orbits at an altitude of 20 000 km has an average radius of 6.3 km. These are the dimensions expected of large Planet X satellite cores, not planetary central cores, which suggest that this may be what they are, all the larger Planet X planetary central cores that once affected Mars having moved on due to its outer cores having become energy depleted (see Article 1083: Planet X destroyed Mars and is now destroying Earth) [2].

Planet X planetary central cores are expected to have a radius of between 400 km and 2400 km, as the largest in the Solar System, Mercury, has a radius of 2400 km, and two others, observed around the Sun in the 1800s, seemed to have radii of 391 km and 938 km (see Article 1028: Earth's new moons and size of Planet X objects observed in 1800s) [3]. In addition, an estimate of the size of one of these objects, which are new earth moons and still come down to a very close distance to the earth's surface yielded a radius of 1146 km (712 miles) (see Article 1011: Planet X cataclysm: size and closeness of the objects responsible) [4]. Then, satellite cores would be expected to be at least 100 times smaller than central cores, as star central cores eject cores which then turn into planets, and since the Sun is 100 times larger than the earth, we would expect satellite cores to be at least 100 times smaller than central cores.

Since a Planet X system may have some 10 000 to 100 000 satellite cores in it, it is likely that the earth is now surrounded with possibly millions of small satellite cores and that therefore the true debris in orbit around the earth is not man-made at all, but rather natural Planet X cores, which have gone through the usual energy absorption process and have reached a state of energy equilibrium, and are thus now in near circular orbits. The man-made debris field is thus likely to be an attempt to cover-up what is really in orbit around the earth, i.e. Planet X.

In conclusion, the idea that earth is surrounded by a man-made cloud of debris is illogical and suggests an attempt to cover up Planet X as the true source of the debris field, which would be made up of Planet X satellite cores that have reached energy equilibrium, with the earth's core system, and are thus now in near circular orbits.

References:

[1]   Albers, C. (2019). Article 1040: Aliens have filled humanity's knowledge base with lies.
[2]   Albers, C. (2019). Article 1083: Planet X destroyed Mars and is now destroying Earth.
[3]   Albers, C. (2019). Article 1028: Earth's new moons and size of Planet X objects observed in 1800s.
[4]   Albers, C. (2019). Article 1011: Planet X cataclysm: size and closeness of the objects responsible.

Chapter 10

1110. Planet X in the sky over Northern California: gravity connections

Figure 1 below shows a photograph of the cloud formations which appear as a flat plane and are clumpy, which indicates that they are cloud forming on the surface of one of the huge Planet X cores which are coming down to a very close distance to the earth's surface and absorbing energy from the earth's core system which then allows them to create cloud on their surfaces.

Figure 1. Left: This seems to be another example of a large Planet X core or Stellar Core (SC), in the sky, it also has a sparse cloud envelope, a blue surface and a cloud spout, connected to its lowest point, as evidenced by the circular region. The overall curvature of the surface can be discerned. **Right:** Another example of another large blue SC in the sky, but this time the hole at the bottom of the object, where its spout cloud connects is not circular in shape. Notice that they all have a different tone of blue, suggesting they all absorb slightly different frequencies of light (see Book 12: Planet X: The Greatest Cover Up in World History) [1].

Figure 10.2. Planet X cores or Stellar Cores (SCs) inside the earth's atmosphere. The smaller is a satellite core to a planetary central core, which the larger most likely is, as it would have a radius of between 500 km and 1000 km.

The larger cores create clumpy and often sparse cloud envelopes on their surfaces with holes in between the clumps of cloud through which it is possible to see the blue light emitting surface. Their cloud envelopes are most likely sparse because they are not able to absorb energy as fast as the smaller cores. But the fact that the matter creation, on their surfaces, is in the form of water vapor, i.e. cloud, indicates that they are absorbing energy from water creating cores in the earth's core system. These larger objects also do not come to, as low an altitude as the smaller cores, which indicates that they are more strongly repelled by the earth's surface than the smaller cores.

Figure 10.3. The same huge central core as appears in figure and the one which made me realize that these objects are coming in very close to the earth's surface as it clearly shows a solid blue surface in the sky. The sparseness of the cloud envelope and the size of the cloud spout at the clearing or gravitational connection point suggests a slow absorption of energy through the connection it is making to water creating cores below the surface of the earth (see Article 1029: Planet X as new moons orbiting the earth: the irrefutable evidence) [2].

The gravitational interaction is made up of two opposing forces, one attractive and the other repulsive, the attractive is stronger than the repulsive until two cores approach each other to a certain minimum distance at which time the repulsive becomes stronger. And the two objects then repel each other as if they were two north poles of a magnet. Gravity is also dependent on energy, so the higher the energy state of an object, the stronger the gravitational interaction, but both the attractive and repulsive get stronger, so that the distance at which the object is repelled increases. This causes its orbit to get further away from the object, from which it is absorbing energy. Gravitational connections are about energy transfer, when an object is orbiting another, it is absorbing energy from the object it is orbiting.

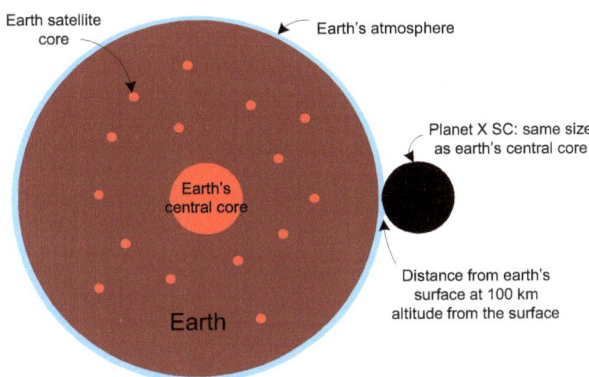

Figure 10.4. Illustration of how closely the huge Planet X cores creating cloud on their surfaces are approaching the surface of the earth: Only an object that is extremely energy depleted would be able to approach earth as closely as the Planet X central core in figure 1.

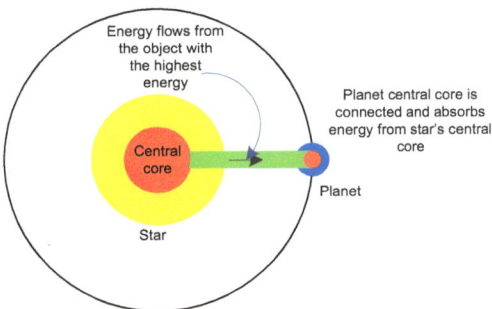

Figure 10.5. Both stars and planets have core systems and a central core. Planets form when a star's central core ejects a satellite core outside of its body. The ejected core becomes the central core of the planet. It remains gravitationally connected to the star and absorbs energy from it. Energy absorption is done through matter and so the central core has to create enough matter through which it then absorbs energy. Thus, a planet maintains energy equilibrium, so that it can maintain its circular orbit, by creating matter with which to absorb energy, so that it can absorb enough energy to continue to connect to the star. If it is not able to maintain energy equilibrium, at a certain distance to the planet, it will move to a closer orbit. If it gains too much energy for the orbit it is in, it will be repelled and move to a higher orbit.

Since it takes matter for a planet and a star to remain connected, the two will exchange matter, or at least the planet will continually drop material it creates, through the connection, toward the star and pull matter from the star towards itself. Energy flows through the physical connection the matter creates, the matter connection is thus an energy conduit. The planet will use matter, which it creates as a cloud envelope, i.e. its atmosphere, and so will the star.

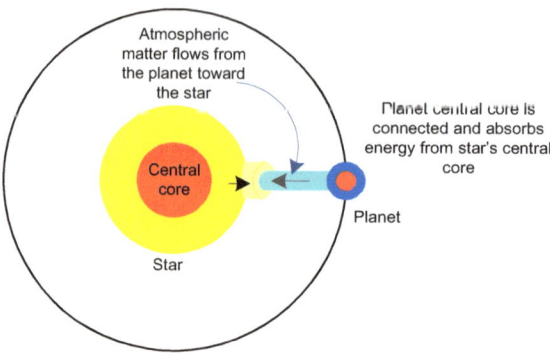

Figure 10.6. Planets and stars will continuously exchange matter; matter from the star's surface will not always reach the planet, but matter from the planet must reach the star, in order for a good energy conduit to be established. Thus, atmospheric matter from the earth must continuously flow toward the Sun.

This understanding of how gravity works comes from observing Planet X cores making connections with the earth. Planet X cores drop cloud or rain onto the surface of the earth, in order to create an energy conduit, through which to absorb energy and the denser the matter conduit the greater the rate of energy absorption. Energy can be absorbed directly through the atmosphere of a planet or a star's very tenuous outer atmosphere; the heliosphere, in the case of the Sun, but the rate of energy transfer is very slow because of its low density. This may however be the only way that the energy depleted Planet X cores or Stellar Cores (SCs) start absorbing energy from the Solar System.

Figure 10.7. A small Planet X core or SC drops cloud, material it has created onto the surface of the earth in the gravitational connection region, which then becomes a conduit through which energy flows towards it (see Article 695: Planet X creates water: it started at the Flood) [3].

Figure 10.8. When the SC has reached a higher energy state, it is able to more quickly make a connection and is also able to exchange matter, i.e. matter from the surface of the earth is pulled upwards toward it, which creates a better energy conduit, so that energy transfers event faster. These objects thus create tornado effects: they can pull roofs off houses and cause buildings to be obliterated due to differences in density in the materials inside the gravitational connection region.

In conclusion, huge Planet X cores create a thin layer of cloud on their surfaces, which suggests that they are connecting to earth water creating cores under the surface of the earth. Their very close approach to the surface of the earth indicates that they are still extremely energy depleted at this stage. Gravitational connections are done with matter created by the object connecting and absorbing energy from another, the physical connection between the two becomes an energy conduit and shows that matter from the earth's atmosphere must be continuously flowing toward the Sun in the connection region between the Sun and the Earth.

References:

[1]   Albers, C. and C'one, S. (2019). Book 12: Planet X: The Greatest Cover Up in World History.
[2]   Albers, C. (2019). Article 1029: Planet X as new moons orbiting the earth: the irrefutable evidence.
[3]   Albers, C. (2019). Article 695: Planet X creates water: it started at the Flood.

Chapter 11

1111. Planet X orbits and the Moon creates its own atmosphere

In Article 1110: Planet X in the sky: gravity connections [1], I discussed how these objects must be extremely energy depleted in order to be able to approach the surface of the earth to such a close distance and that they must be connecting to cores close to the surface of the earth, i.e. water creating cores. I then discussed how gravitational connections are about energy transfer and how Planet X observations lead to the understanding of how gravitational interactions between planets and stars work. In this article, which is basically a continuation of Article 1110, I will go on to discuss the orbits that the SCs will tend to follow as they absorb energy from the sun's or earth's core system, until they finally reach energy equilibrium and thus become like the Moon, which is now in a close to circular orbit. Figure 1 below shows another photograph sent in by Beth of the surface of the same object appearing in Article 1110.

Figure 1. An apparent circular hole is actually the surface of a huge Planet X core in the sky which is creating a huge surface covered in layer of cloud. The blue clearing is actually a connection point and the stringy cloud attached to it is cloud spout material (see Book 12: Planet X: The Greatest Cover Up in World History) [2].

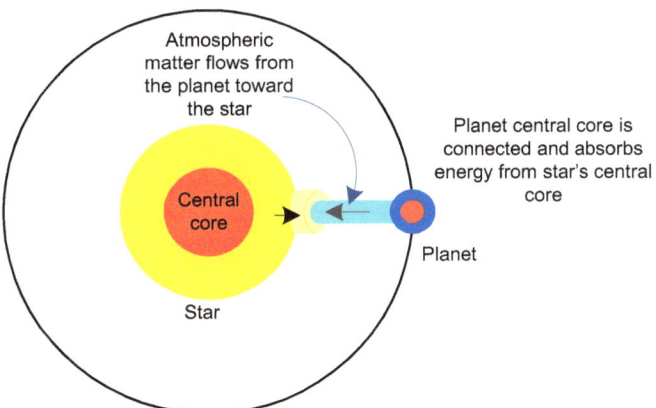

Figure 11.2. Planets and stars will continuously exchange matter; matter from the star's surface will not always reach the planet, but matter from the planet must reach the star, in order for a good energy conduit to be established. Thus atmospheric matter from the earth must continuously flow toward the Sun [1].

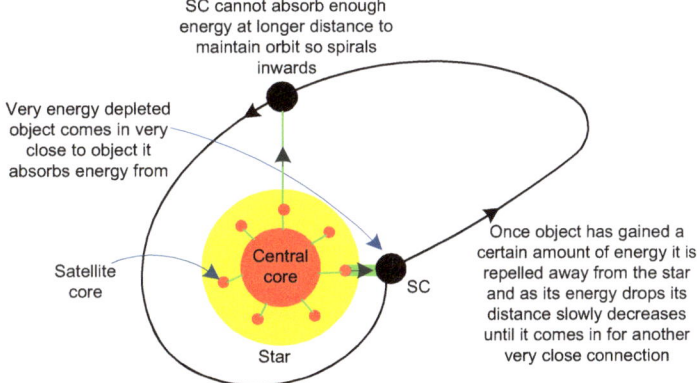

Figure 11.3. Illustration of the orbit followed by a SC when it first arrives, at the object that becomes its host: It comes as close as it can, to the surface of the host and makes a gravitational connection, with the cores inside the body of the host that its gravitational connection is able to reach, and then absorbs energy from those cores it can reach, through the densest matter connection, available, in the connection region. For the largest SCs and the earth, this may just be atmosphere and water, on the surface of the planet. Once a certain amount of energy is absorbed, the SC will be repelled and move away from the host. As it moves away, the repelling force decreases and the attractive force increases, once the attractive force is strong enough it will come back towards the host following a spiraling in orbit.

The SC will create matter with the energy it has gained, and is gaining, in response to the gravitational connection, and as its energy status drops, it gets closer to the host, until it reaches the minimum distance, where it is able to absorb the most energy, in the smallest amount of time. It should gradually be able to connect to cores, deeper within the earth, from which it will obtain a faster energy transfer and also its minimum approach distance, or perihelion of its orbit, will increase.

The fact that SCs create matter in response to a gravitational connection suggests that the Moon will create some type of atmosphere and that some of this material is continuously moving toward the earth. There is therefore most likely always some type of atmosphere between the earth and the Moon.

In conclusion, Planet X cores, which are coming very close to the earth's surface are extremely energy depleted, but the perihelion of their orbit will gradually increase, until they eventually orbit the earth in a close to circular orbit. As all objects orbiting another object must create matter, in response to a gravitational connection, Earth's original Moon must create its own atmosphere, with which it connects and absorbs energy from the earth's core.

References:

[1]   Albers, C. (2019). Article 1110: Planet X in the sky over Northern California: gravity connections.
[2]   Albers, C. and C'one, S. (2019). Book 12: Planet X: The Greatest Cover Up in World History.

Chapter 12

1112. The earth's atmosphere creates daylight on one side: the reason

In Article 1110: Planet X in the sky over Northern California: gravity connections [1], I showed that earth creates matter with the energy it obtains from the Sun's central core, which then flows out from the side of the earth facing the Sun, in the region of gravitational connection, between the two objects.

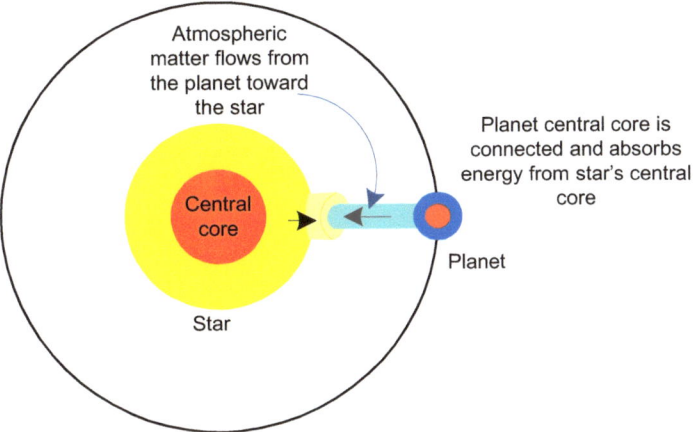

Figure 12.1. Planets and stars will continuously exchange matter; matter from the star's surface will not always reach the planet but matter from the planet must reach the star in order for a good energy conduit to be established. Thus, atmospheric matter from the earth must continuously flow toward the Sun [1].

This also explains why earth's atmosphere produces daylight, on the side facing the Sun (see Article 888: The Sun is gone: Daylight comes from earth's core) [2]. Energy flows through the daylight side of the earth, in order to create matter, in response to the gravitational connection to the Sun. The matter flows toward the Sun and establishes an energy conduit, through which energy can then flow from the Sun's central core to the earth's central core. The energy flowing out from the earth's core, which produces daylight seems to come from a certain type of core, inside the body of the earth, which seems to be close to the surface of the planet, and which is subject to energy depletion, as it has been observed that the earth's atmosphere becomes dark when a Planet X system, comes in over a certain region, and makes a gravitational connections, which extract a large amount of energy (see Article 1072: Black rain falls and darkness occurs in São Paulo Brazil due to Planet X) [3].

Figure 12.2. A tornado causes the atmosphere to darken for a few seconds to near nighttime darkness. This occurred suddenly without any new clouds possibly moving in, suggesting that the energy, which allows earth's atmosphere to produce daylight was suddenly not there anymore (see Article 923: Planet X tornado causes the atmosphere to darken in Chile) [4].

Energy depletion of the cores that produce daylight is the most likely reason why the Sun is no longer emitting visible light (see Article 1103: Hercolubus and Earth's new red moons: Planet X in a Polar Orbit) [5].

In conclusion, the understanding of how cores respond to energy connections which allow them to absorb energy, which is by creating matter which then flows through the connection to the parent or host body, and allows faster energy transfer, leads to the understanding that it is precisely this effect which causes earth's atmosphere to emit light and thus produce daylight on the side facing the Sun.

References:

[1]   Albers, C. (2019). Article 1110: Planet X in the sky over Northern California: gravity connections.
[2]   Albers, C. (2019). Article 888: The Sun is gone: Daylight comes from earth's core.
[3]   Albers, C. (2019). Article 1072: Black rain falls and darkness occurs in São Paulo Brazil due to Planet X.
[4]   Albers, C. (2019). Article 923: Planet X tornado causes the atmosphere to darken in Chile.
[5]   Albers, C. (2019). Article 1103: Hercolubus and Earth's new red moons: Planet X in a Polar Orbit.

Chapter 13

1049. Electrons are photons moving at less than the speed of light

Figure 1 below shows a screenshot from a very interesting film on matter waves, from 1953. The viewer is clearly shown that particles behave like waves. This is known as the wave particle duality of light and matter, light is made of particles called photons and particles, such as electrons, exhibit wave properties like diffraction. This idea was first proposed by de Broglie in 1923 and it was a few years later experimentally verified to be correct.

Figure 13.1. Screenshot from a film on matter waves (**Source**: YouTube Video by PeriscopeFilm, 11 August 2019, entitled: MATTER WAVES & PHYSICS BELL LABS FILM w/ ALAN HOLDEN & LESTER GERMER SCIENCE EXPERIMENTS 19534) [1]

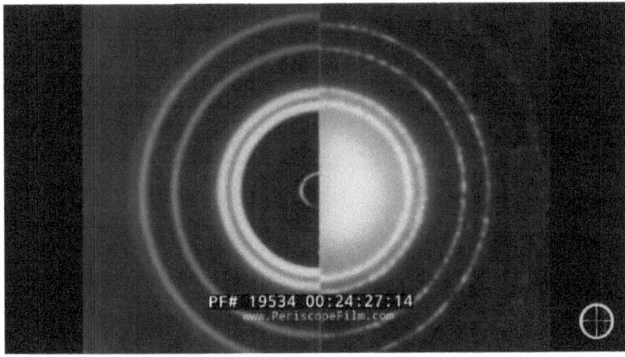

Figure 13.2. Diffraction patterns obtained by passing x-rays (left) and a beam of electrons through a metal foil. The two diffraction patterns are very similar. This is because the electrons and x-rays have the

same wavelength. The obvious difference is that the first bright ring in the x-ray diffraction pattern is missing in the electron diffraction pattern and that the x-ray bright rings are of higher intensity (thicker).

This is very interesting but what I have discovered through Planet X observations goes much further. What I discovered is that the gravitational interaction is dependent on energy and this energy is light or photons. This energy exists as photons within matter and explains why the temperature of material increases when it absorbs radiation.

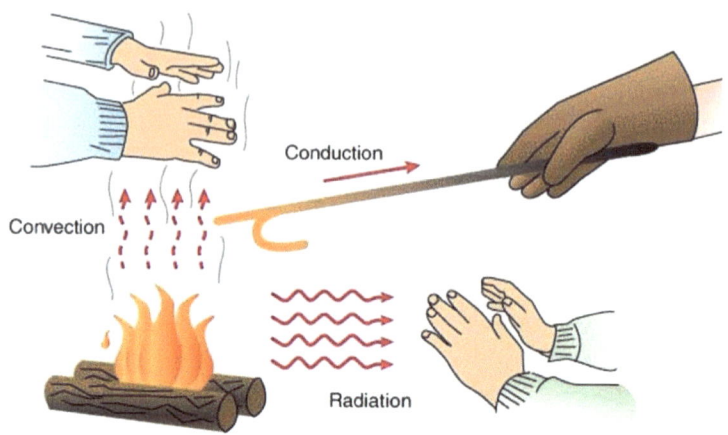

Figure 13.3. The fire emits radiation which when it is absorbed by a person's hands causes their temperature to increase.

But heat is movement of particles or kinetic energy and thus kinetic energy is dependent on the number of photons within matter. Then, since electrons can be accelerated by moving across a potential difference, without absorbing any photons, it can be deduced that the electron itself has to be a photon or a collection of photons. This agrees with de Broglie's idea that electrons have wave properties but it goes further because it means that particles are actually photons. But these photons, which behave like electrons, are moving at speeds less than the speed of light and exhibiting a property called mass. Mass is associated to inertia, electrons suffer from a quality called inertia, i.e. they resist moving when a force acts on them, to a degree which is dependent on their mass.

Thus, everything in the universe is made of light or of photons, all energy is photons, all light is photons and all particles are photons, or we could say that all energy and particles are light. Electrons are actually a manifestation of an energy differential or potential energy difference, which causes energy to be transferred from one point to another, electrons are energy being transferred and when the potential equalizes and there is therefore no need to transfer energy anymore, electrons cease to exist. Thus, electrons arise out of matter which is made of a photon fluid or plasma (see Article 655: The destruction of the Planet X planets and the electron) [2].

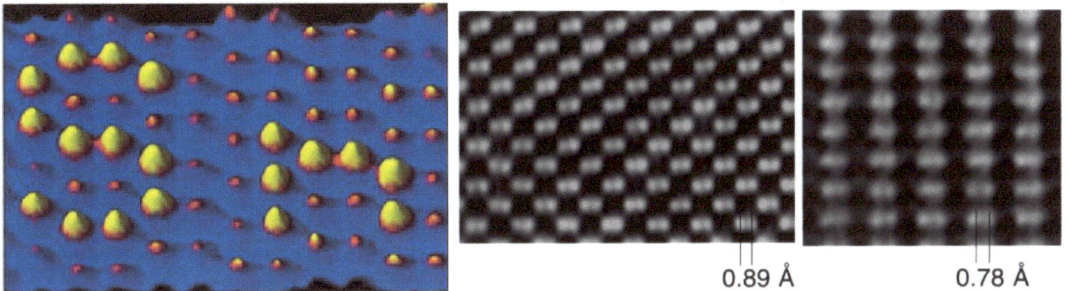

Figure 13.4. Atoms seem to have a regular shape, but they are not solid, they are made of light and can flow, and are thus, made of liquid plasma. Pieces of these atoms, which we call electrons will emerge from inside them, and flow in a certain direction, in response to an electric potential, i.e. to energy which is to be transported from one region to another, across the material. **Left:** Atoms look like blobs of plasma. **Right:** the image on the right is at a higher magnification and we can see a swirling flow of material connecting atoms indicative of a liquid plasma or photon material (see Article 655: The destruction of the Planet X planets and the electron) [2].

In conclusion, a film on matter waves from 1953 shows how all particles have wave properties and thus partially illustrates what I have found through Planet X observations: everything in the universe is made of light, all energy and all particles are made of photons or light.

Acknowledgements:

Thank you to Rik Watley who pointed out the YouTube Video showing the Matter Waves film in a comment.

References:

[1] Video by PeriscopeFilm, 11 August 2019, entitled: MATTER WAVES & PHYSICS BELL LABS FILM w/ ALAN HOLDEN & LESTER GERMER SCIENCE EXPERIMENTS 19534. https://www.youtube.com/watch?v=cx-7FAEV6AM

[2] Albers, C. (2019). Article 655: The destruction of the Planet X planets and the electron.

Chapter 14

655. The destruction of the Planet X planets and the electron

In Article 645: Moon in atmosphere due to Planet X and electrons are light [1], I showed that since electrons are supposedly carrying gravitational energy, and yet that their energy is in a kinetic form, i.e. associated to their speed, that electrons have to actually be photons, photons with a charge, where their charge distinguishes them from other particles, which have opposite charge, i.e. protons. Then protons would be photons, with positive charge and electrons, photons with negative charge. Since, according to my theory of gravity protons and electrons form when photons split due to passing through a high electric field, it makes sense that the photon splits into two parts, with different qualities.

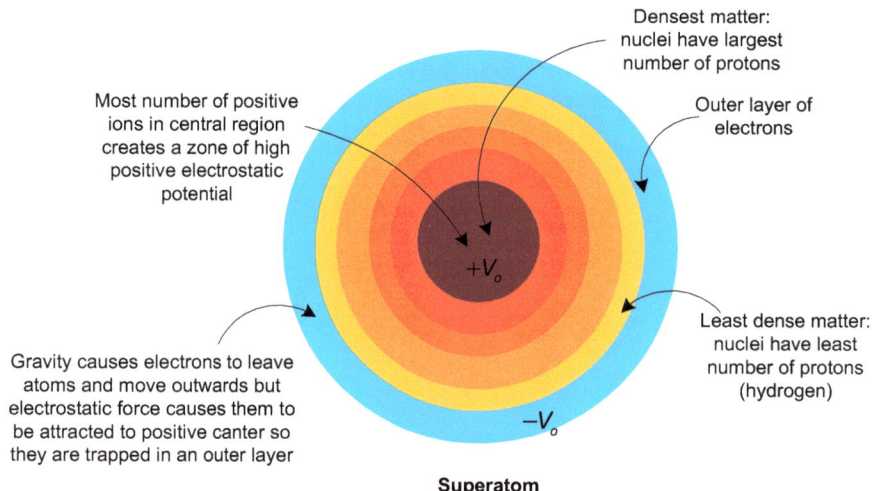

Figure 14.1. The core of a planet has the highest amount of gravitational energy and it causes electrons to flow out toward the edges of the celestial object and transmit energy to atoms in those outer layers.

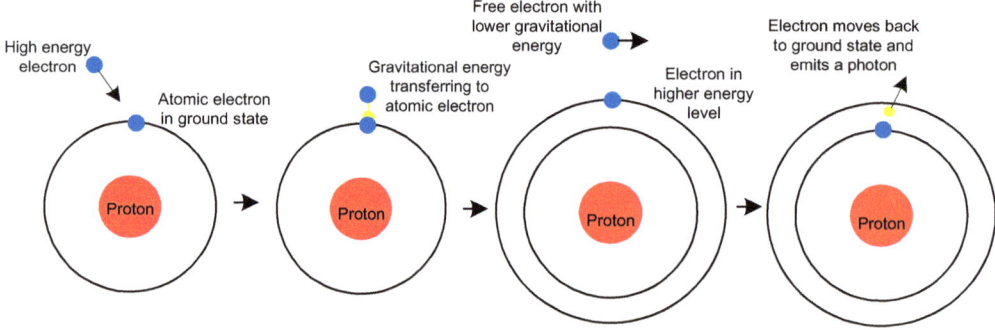

Figure 14.2. Gravitational energy coming from the core is transmitted to atoms through free electrons interacting with atomic electrons and transmitting a piece of themselves to the atomic electrons, which causes the free electron's speed to drop because an electron's energy is associated to its speed.

Thus, gravitational energy is a photon and it manifests as speed. But if the electron is a photon, then an electron is also gravitational energy, which manifests as speed. So what happens if the speed of an electron drops to zero? The electron will cease to exist. This means that electrons are not particles in the sense that they are indestructible tiny solid objects, which once formed never cease to exist. Electrons are actually a manifestation of an energy transfer and appear, from within matter, whenever there is energy to transfer, and disappear, when there isn't. Photons do the same thing across material, which allow the free flow of photons: the vacuum of space and the atmospheres of planets and stars, which are gaseous in nature, and transparent to light.

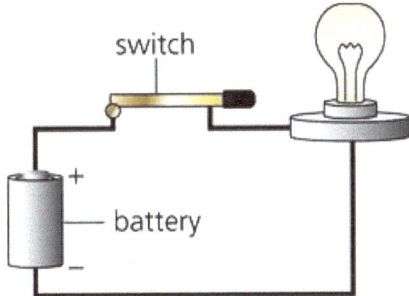

Figure 14.3. When the switch is closed energy flows from one terminal of the battery to the other the energy is carried by electrons, which appear when the switch is closed and disappear when it is opened again, because it stops the flow of energy. Thus, electrons appear from within atoms only when there is energy to transfer.

This means that matter is a fluid, a photon fluid, since all matter is photons. This photon fluid has higher concentrations within it, called atoms, and from which pieces of the photon fluid can separate and flow in a certain direction in response to an energy flow across the material.

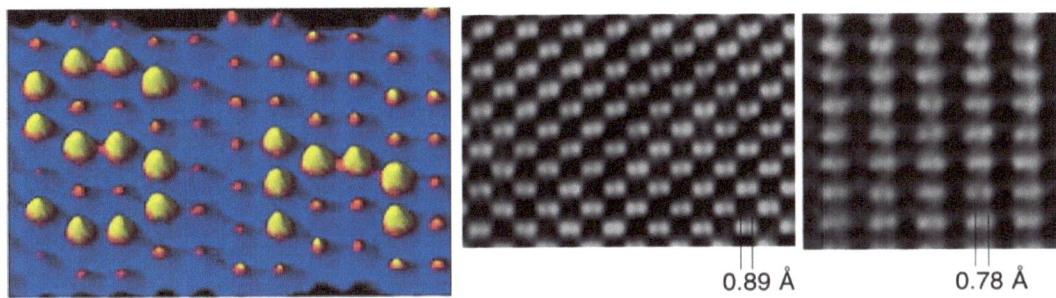

Figure 14.4. Atoms seem to have a regular shape, but they are not solid, they are made of light and can flow, and are thus, made of liquid plasma, just like the cores of celestial objects. So, once again, the universe is the same on the microscopic as well on the macroscopic level. Pieces of these atoms, which we call electrons will emerge from inside them, and flow in a certain direction, in response to an electric potential, i.e. to energy which is to be transported from one region to another, across the material. **Left:** Atoms look like blobs of plasma. **Right:** the image on the right is at a higher magnification and we can see a swirling flow of material connecting atoms indicative of a liquid plasma or photon material.

Thus, when the Planet X planets lost their gravitational energy due to losing their energy connection to the super galactic core, the flow of energy stopped and thus all electrons disappeared. But they did not cease to exist, they were just left with much less mass, which serves as an indicator of gravitational energy, which suggests that all celestial objects have an inherent gravitational energy, when they are created and that the rest, which must be a very large proportion continuously flows into its core, through the connection it has to its parent.

Energy always flows from a point that has more energy to a point that has less energy, and thus flows from the core of a celestial object outwards, and from a larger core toward a smaller core.

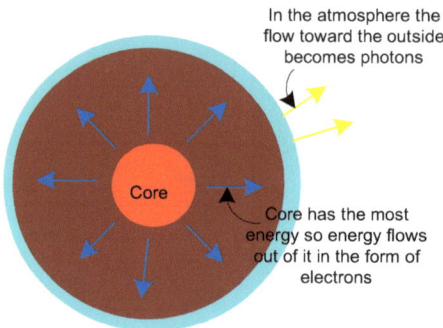

Figure 14.5. Energy transfers from the point with the highest energy, in a celestial object, i.e. the core to the outside, in the form of electrons, through solid matter, and in the form of photons, through the atmosphere and out into space.

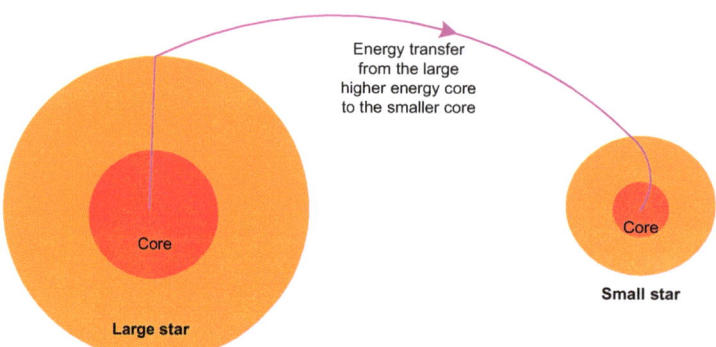

Figure 14.6. Energy transfers from the core of a large star, to the core of a smaller star, or planet, that it has created, because the connection is established at the creation event, and then energy, continuously flows from the high energy point to the lower energy point.

The fact that electrons can just disappear when no energy flows, shows that charge is not conserved, charge is a concept that can be used to keep track of energy flow, but it is not a basic concept, energy flow is. And what differentiates the proton, from the electron, may simply be the direction of spin, so that charge is just a different spin direction. Another thing, which differentiates the two different parts of a photon, i.e. the proton and the electron, is that the electron responds very quickly to an accelerating potential, whilst the proton is slow to respond, and this difference in response time is what determines another characteristic called mass. The proton has much more of it than the electron.

Now, in order to create levitation of an object in the earth's atmosphere, we would have to decrease its gravitational energy, but since energy would tend to flow into it as we force its evaporation, we would have to continuously decrease its energy. The best way to do that seems to be by spinning electric charges in a magnetic field. To set up the charges we would need a high electric potential and then we would need to create a particle accelerator type conduit to accelerate the charges in. Accelerated charged particles emit light or photons and thus decrease their gravitational energy.

In conclusion, electrons are photons, i.e. liquid plasma, which exists in the form of tiny blobs, inside what we perceive as matter. Electrons appear from within atoms, when there is energy to be transported from a region of high energy to another region of lower energy.

References:

[1]   Albers, C. (2019). Article 645: Moon in atmosphere due to Planet X and electrons are light.

Chapter 15

645. Moon in atmosphere due to Planet X and electrons are light

It has been recently discovered that the earth's outer atmosphere, which is made of hydrogen atoms, stretches further than had been originally thought, according to 20 year old observations done by the SOHO spacecraft. This part of the earth's outer atmosphere is called the geocorona and it was previously thought that it went out to 15.5 earth radii, i.e. $15.50r_E$, but according to the SOHO data it goes out to $100r_E$. Since the moon orbits the earth at about $60r_E$, the moon would therefore be inside the earth's geocorona and thus technically inside the earth's atmosphere. The density of matter decreases in a celestial object, as we move from the center outwards, so the geocorona is the least dense layer, and so tenuous that there are less than 100 atoms per cubic centimeter. In the atmosphere, close to the surface, there are typically 10^{19} atoms, per cubic centimeter, i.e. 10^{17} order of magnitude more (1 followed by 17 zeros factor difference). Thus, the geocorona is extremely tenuous and therefore a very good vacuum, by laboratory standards, which can only be obtained with the very best laboratory vacuum pumps (10^{-14} torr). However, the geocorona emits ultraviolet light due to collisions between high energy electrons and hydrogen atoms, which excites the atoms and when they return to the ground state, a photon is emitted. High energy electrons have high velocity and thus high kinetic energy and transmit this kinetic energy to the atomic electrons through collisions.

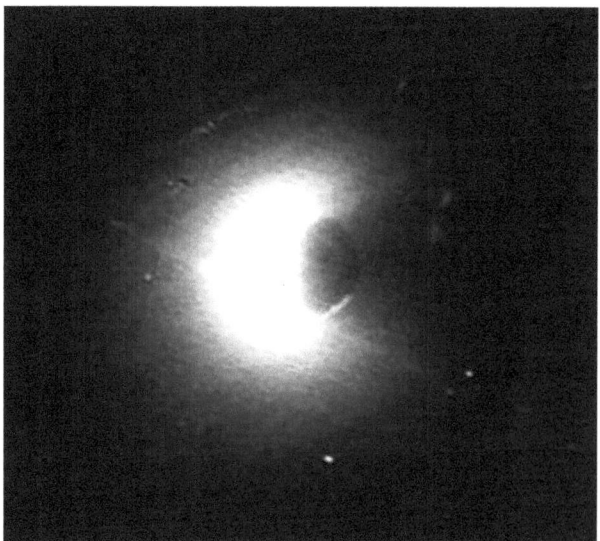

Figure 15.1. The earth's geocorona from a 1972 photograph. In this photograph, the geocorona seems to go out no further than $5r_E$. So if the geocorona was found to go out to $100r_E$, 20 years ago, or in 1999, this means that the geocorona grew to a huge size (by a factor of 20) between 1972 and 1999.

Scattering of solar radiation is usually the explanation, which is given for the light emitted by geocorona but it is impossible that scattering is what is occurring here, only excitation of the hydrogen atom due to high energy electrons being present, at such altitudes, can explain the emission of the particular wavelength of light emitted, which is due to an energy transition in the hydrogen atom, called the Lyman alpha transition. This energy transition leads to the emission of a photon of wavelength 1215.67 angstroms. The high energy electrons that start the process, which leads to the emission of light, at this wavelength, are at this altitude because of the earth's gravitational energy. The earth's core repels electrons with the most gravitational energy the furthest distance from the core, as a result of the charge separation part of the gravitational interaction.

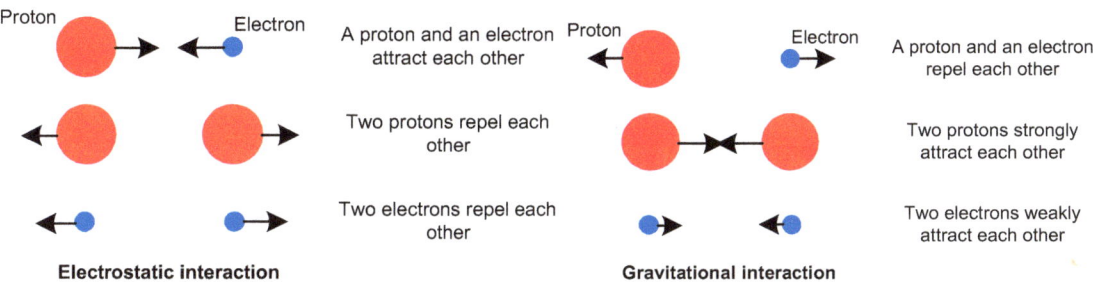

Figure 15.2. The electrostatic and the gravitational interactions between protons and electrons (see Book 3: Planet X Revealed Gravity and Light) [1].

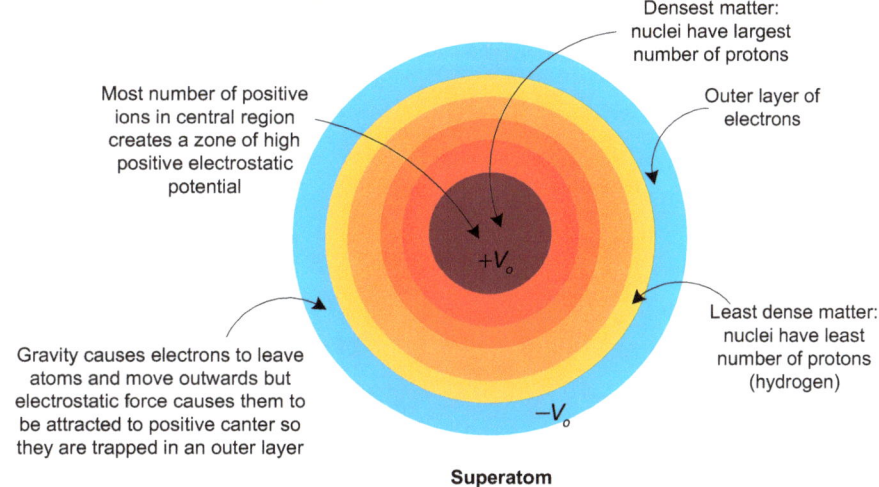

Figure 15.3. The layers making up a celestial object all come from the core, which creates matter, the density of the layers decreases, as we move toward the surface, and the gravitational potential decreases as well. It is the charge separation part of the gravitational interaction, which causes the outer least dense layer, or atmosphere, to be negatively charged; the surface is neutral and the interior is positively charged.

Since the highest gravitational energy electrons have the highest gravitational energy, this suggests that it is the gravitational energy inside the electron, which gives them kinetic energy and thus velocity, and that therefore the collision leads to a transfer of gravitational energy (photons) from one electron to the other. This must be due to the gravitational interaction between electrons, i.e. the third part of the gravitational interaction, shown in figure 2. Thus, kinetic energy is a manifestation of the gravitational energy between electrons.

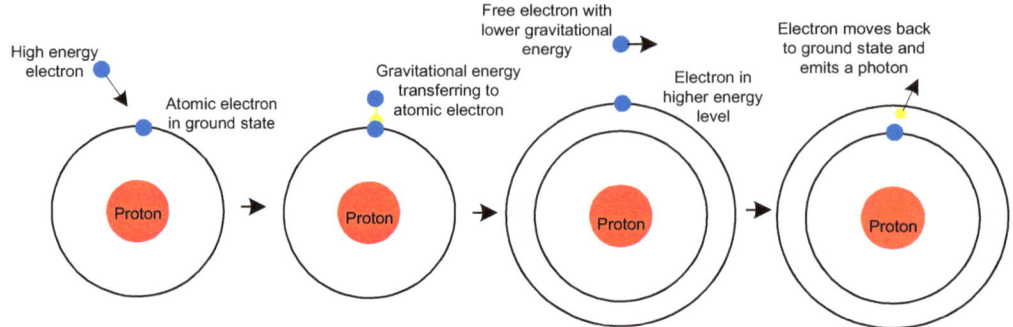

Figure 15.4. Gravitational energy transfers between electrons when they collide, leading to the emission of a photon by the atomic electron. The energy transfer between electrons is as a result of the electron-electron gravitational interaction.

Thus, gravitational energy is the same as kinetic energy. But then it is possible to get an electron to gain gravitational energy by placing an electric potential across it. An electric potential is an electric field and if an electric field can directly affect the gravitational energy of an electron, so that it slows down or speeds up, without any photon entering it, the only conclusion that can be made is that an electron is just a photon or a collection of electrons, with charge. It is not that the photon is inside the electron, but that the electron is made of photon, which has a charge and mass. In the same way, a proton must also be a photon (or a collection of photons) with charge and mass. The charge appears, when the photon splits into two parts as it moves through a region of high electric field, and so does the mass. Thus, an electron-electron transfer of energy is the electron transferring part of itself to another electron.

The earth's geocorona having greatly increased in size, suggests that the earth's electrons are reaching further out from the earth's core. This would require the earth to now have more gravitational energy than before, which does not seem possible since the Planet X System objects are coming in, in greater numbers and absorbing gravitational energy, which suggests that the electrons are not getting to that altitude by themselves, they are being carried by the hydrogen atoms, which thus become negative ions as they move upwards from lower altitudes toward the geocorona. In order for hydrogen to be found at higher altitudes, there has to be a lot more hydrogen in the earth's atmosphere than before. Since the earth is absorbing Planet X debris which are the outer layers of the broken up planets and stars that make up the system, the extra hydrogen is most likely coming from this absorbed debris. Planet X atmosphere comes into the earth's atmosphere in a frozen form and is known as snow. Snow is not just water as it contains other elements, which sublimate to a gas, and must therefore be Planet X atmosphere. It sometimes contains chemicals, from the atmosphere of the planets, which turns black, when burned, which clearly shows that it cannot be just water (see Article 615: Snow turning black when burnt has alien chemical compounds) [2].

In conclusion, the earth's geocorona having greatly expanded since 1972 indicates that the earth is absorbing a lot of Planet X atmosphere, which comes in as snow. The energy transfer leading to the geocorona's emission of ultraviolet light suggest that kinetic energy is the same as gravitational energy and that electrons do not contain photons, electrons are photons with a negative electric charge, and electrons transfer energy, to other electrons, by transferring a part of themselves to other electrons.

References:

[1] Albers, C. and C'one, S. (2018). Book 3: Planet X Revealed Gravity and Light.
[2] Albers, C. (2019). Article 615: Snow turning black when burnt has alien chemical compounds.

Chapter 16

1050. The wave nature of particles comes from the gravitational interaction

In Article 1049: Electrons are photons moving at less than the speed of light [1], I wrote about matter waves, which causes particles, such as electrons, to exhibit a wave like nature, so that they can be diffracted just like photons can and in fact since electrons have a wavelength, which falls within the wavelength range of x-rays, they also have very similar diffraction patterns to x-rays when passing through a metal foil, where the crystal structure of the metals acts as a diffraction grating.

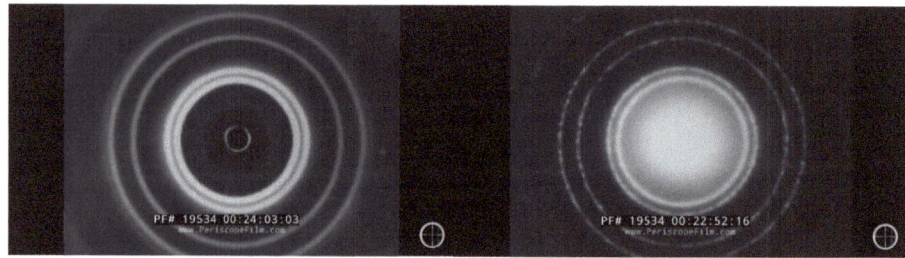

Figure 16.1. Diffraction pattern for x-rays (left) and electrons (right) obtained when these pass through the same metal foil.

However, single electrons are not diffracted and if single particles are projected toward a double slit arrangement, they will pass through one of the slits and no interference pattern will be produced. It is only when a beam of electrons is projected toward a double slit or diffraction grating that a diffraction pattern will appear. This suggests that the wave nature of electrons and indeed all particles is due to the interaction between them, when there are a large number of these particles moving together. The same thing happens at other matter scales. Light or photons have a wave nature, which manifests as the production of diffraction pattern, when a large number of photons pass through 2 slits of a size comparable to the wavelength of the photons.

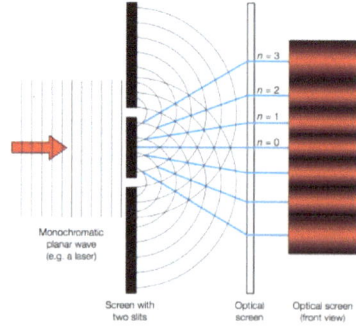

Figure 16.2. Diffraction of light showing that photons have wave properties: When light passes through a slit, light spreads out and thus the slit acts as a source of light. Any light reflecting off the sides of the slits will be diffracted as well, it will move backwards toward the source of the light.

Vortices form at all levels in the universe, a galaxy's spiral arms are a vortex and Planet X cores, inside the earth's atmosphere, form gravitational connections, which give rise to vortices. These can be clearly seen in the case of waterspouts and tornadoes. These vortices form because of the cores inside their cloud envelopes are sources of gravity and therefore a vortex is a wave or a diffraction pattern, which arises as a result of the gravitational interaction.

Figure 16.3. Left: A waterspout is a gravitational vortex. Water can be seen spiraling upwards around a central space. Water is forced upwards by a gravitational force which pulls it upwards. The source of the gravitational force is a Planet X core inside the cloud. **Right**: A Planet X core inside its cloud envelope, these objects are sources of gravity and create matter, cloud or water, as a result, because gravity is a creative force (see Article 785: Planet X is here but what is it exactly?) [2]

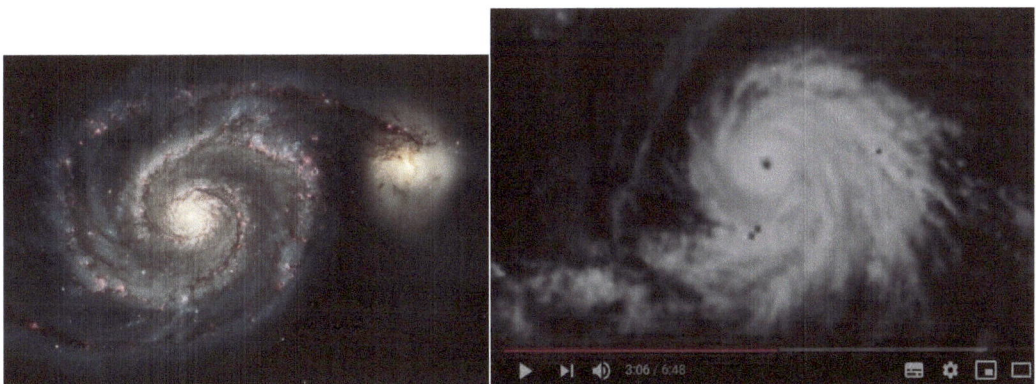

Figure 16.4. Left: The spiral arms of a galaxy are in the form of a flat vortex but a vortex nevertheless because there is a source of gravity and matter inside the nucleus, a galactic core (see Article 126: White Holes instead of Black Holes at the Center of Galaxies) [3]. **Right:** A hurricane inside the earth's atmosphere, containing a core in the nucleus of the hurricane, which creates cloud that adopts the same vortex formation as that of a galaxy, because the core is a source of gravity, just like the core in the galactic nucleus. Several satellite cores, which create weather inside the earth's atmosphere, can be seen in this satellite image. Their cloud envelopes seem to have holes at the top which allows them to be seen from above (see Article 978: Planet X observed in earth satellite images) [4].

Thus, the appearance of a vortex signals the presence of a gravitational source but when a large amount of water moves through a small hole, a water vortex also forms, which suggests that the hole acts as a source of gravity, without actually being one, just like a slit in a light diffraction experiment acts as a source of light and therefore a vortex is a wave effect, it is a diffraction effect, it shows that a large quantity of water has a wavelike nature and all matter when congregating in large quantities, such as in the case of galaxy, develops a wave like nature and that this nature arises as a result of the gravitational interaction.

Single droplets of water falling through a drain will go right through without any vortex or diffraction effects but when the sink if full of droplets of water, a vortex forms showing that the gravitational interaction between the large quantity of water droplets is causing what appears to be the wave nature of water to manifest. But the wave effects actually arise as a result of the gravitational interaction between the large numbers of water droplets.

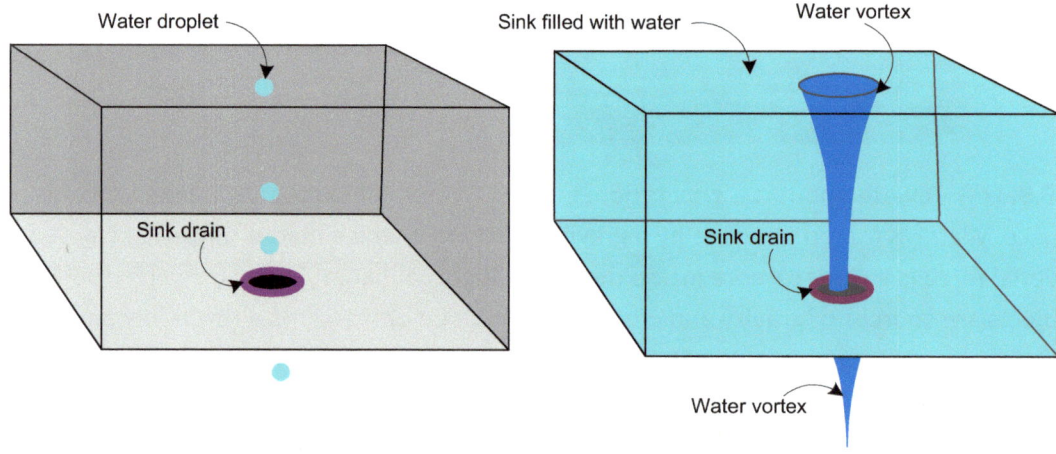

Figure 16.5. A water droplet will fall right through a sink drain without any gravitational vortex effects appearing. But if the sink is filled with water, the vortex appears, and a vortex also forms under the drain. The vortex is widest where the force is strongest, and in the case of the sink, it is strongest at the surface of the water. The water going through the hole continues to fall as a vortex, after it passes the hole, i.e. there are diffraction effects above and below the hole just like in the case of photons moving through a slit.

In conclusion, whether in the case of photons or any other particle, any wave effects, such as diffraction, arises as a result of the gravitational interaction. The wave effects arise whenever there is a large congregation of matter, at all scales; in a sea of photons, in an electron beam, in water droplets as a part of a large quantity of water and in stars making up a galaxy. The wave effect is not in the particles but in the interaction between the particles or water droplets or stars within the galaxy. There is even another larger scale, above stars in a galaxy that is likely to exhibit the same wave effects, due to the gravitational interaction: large groups of galaxies, in the universe, will form galactic vortices due to the gravitational interaction between all the galaxies in each group, as long as it is a very large group of galaxies.

References:

[1]   Albers, C. (2019). Article 1049: Electrons are photons moving at less than the speed of light.
[2]   Albers, C. (2019). Article 785: Planet X is here but what is it exactly?

[3] Albers, C. (2019). Article 126: White Holes instead of Black Holes at the Center of Galaxies.
[4] Albers, C. (2019). Article 978: Planet X observed in earth satellite images.

Chapter 17

1051. Earth emits cosmic radiation at bow shock due to Planet X

In a recent research publication in the Journal Science Advances, on 3 July 2019, researchers: Terry Liu, Vassilis Angelopoulos and San Lu, reported that the earth is emitting relativistic electrons, i.e. electrons travelling close to the speed of light, at the bow shock, which is supposedly the layer of the earth's magnetosphere where the solar wind meets the earth's magnetic environment [1]. These relativistic electrons are a form of cosmic radiation and due to their high speed very disruptive to living tissue, i.e. they are a deadly form of radiation, which would affect any astronauts in spaceships moving through the bow shock or simply moving through space away from it. This is explained in an article, which appears on the Planet X News website entitled: New Source of Space Radiation has been Detected and Earth is Producing it [2].

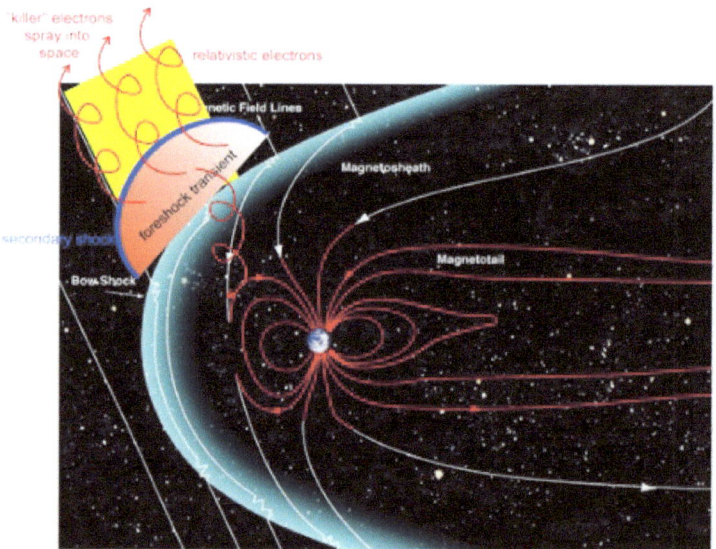

Figure 17.1. Diagram illustrating where the electrons are being generated.

However, electrons only appear when there is energy to be transferred and since they are moving away from the bow shock and thus away from earth, the energy has to be coming from the earth's core and it is moving out toward a region or object, or objects, that are at a much lower energy state than the earth's core is (see Article 1049: Electrons are photons moving at less than the speed of light and Article 655: The destruction of the Planet X planets and the electron) [3, 4].

Electrons, as I explained in the above articles are photons; all of matter and energy in the universe are photons and whenever photons or electrons appear, i.e. are emitted or created, they emerge from within matter, but they do so in order to transfer energy from a point at a high energy level to a point at a lower energy level, and the greater the energy differential the greater the velocity of the electrons is likely to be. This is due to the second law of thermodynamics, which states that any two systems in contact that are not in equilibrium will cause a transfer of energy, from the system at a high energy to the system at a lower energy, until the two systems have reached a state of energy equilibrium. The second law of thermodynamics is the most fundamental law governing the universe because it explains most physical processes in the universe.

Thus, the fact that relativistic, i.e. high kinetic energy electrons are emerging from the bow shock suggests that the bow shock is in contact with objects that are at a very low energy. These objects are of course Planet X since the Planet X planets were destroyed due to energy depletion, causing the outer shell of the planets to break up and thus releasing the cores inside them (see Article 785: Planet X is here but what is it exactly?) [5] Planet X thus comes into the Solar System as core systems with a central core, which would be about the size of a moon, surrounded by small satellite cores, most likely 100 times smaller than the central core, or smaller, and a large amount of debris, all energy depleted (see Article 1029: Planet X as new moons orbiting the earth: the irrefutable evidence) [6]. The emission of cosmic radiation in the form of relativistic electrons suggests therefore that these Planet X core systems are in the space beyond the bow shock or in orbit just beyond the bow shock region.

Figure 17.2. Planet X central core in the sky most likely at its perihelion position, it is likely that its orbit, once it moves outside of the earth's lower atmosphere, will take it beyond the bow shock region of the earth's magnetosphere where it will draw energy from the earth's core in the form of relativistic electrons.

In conclusion, the earth is emitting cosmic radiation in the form of relativistic electrons from the bow shock, which would constitute a great hazard to astronauts inside spacecraft, but the creation of this cosmic radiation can only be due to the presence of Planet X matter, which is energy depleted, in this region of space or beyond, as electrons only emerge from within matter, when there is energy to be transferred.

References:

[1]   Liu, T. et al. (2019). Relativistic electrons generated at Earth's quasi-parallel bow shock. Science Advances, 03 Jul 2019: Vol. 5, no. 7, eaaw1368

[2]   Planet X News website article: New Source of Space Radiation has been Detected and Earth is Producing it. https://www.planetxnews.org/new-source-of-space-radiation-has-been-detected-and-earth-is-producing-it/

[3]   Albers, C. (2019). Article 1049: Electrons are photons moving at less than the speed of light.

[4]   Albers, C. (2019). Article 655: The destruction of the Planet X planets and the electron.

[5]   Albers, C. (2019). Article 785: Planet X is here but what is it exactly?

[6]   Albers, C. (2019). Article 1029: Planet X as new moons orbiting the earth: the irrefutable evidence.

Chapter 18

818. Planet X observations: the electrical universe is made of light

According to mainstream physics, energy is an abstract concept used to understand change in the universe. But Planet X observations reveal, over and over again, that the universe is all about energy. Planet X System Stellar Cores (SCs), the energy depleted core systems of the Planet X star systems, which were destroyed due to energy depletion, come into the Solar System to absorb energy. Energy flows from the sun's or the earth's core system to the SCs. So again these objects reveal that energy and the transfer of energy is what the universe is all about (see Article 785: Planet X is here but what is it exactly?) [1].

Figure 18.1: SDO image in the 171 angstrom wavelength from October 13th 2017 showing a dark Planet X SystemStellar Core making a physical connection with the Sun through which it is absorbing energy from the Sun. The Sun contains inside its body a core system made up of objects like the SC we see here but the ones inside the Sun are fully energized.

The second law of thermodynamics has been shown by Planet X observations to be the most fundamental of concepts regarding all processes in the universe. This law states that all systems will tend to reach a state of equilibrium and this is what is observed. This is the reason why energy flows from one point to another, i.e. in order to establish equilibrium. This is the reason why energy flows from the Sun's core system to a SC. The SC is energy depleted and the Sun is not, so energy flows from one to the other. But what is energy exactly? We can see what energy is when we observe that the absorption of light by a material leads to an energy increase (manifests as a temperature increase), in other words, energy is light. This light is made of photons. Photons are the tiniest particle ever encountered. A 100 W light bulb emits 2×10^{20} every second, that is, 200 quintillion photons every second. The light emitted by this light bulb is made of free photons, which travel at the speed of light.

They are free because they do not deviate from their path, until they are absorbed by matter. In addition, there are always so many photons moving together that it is like a fluid flowing through space. Thus, light is a fluid.

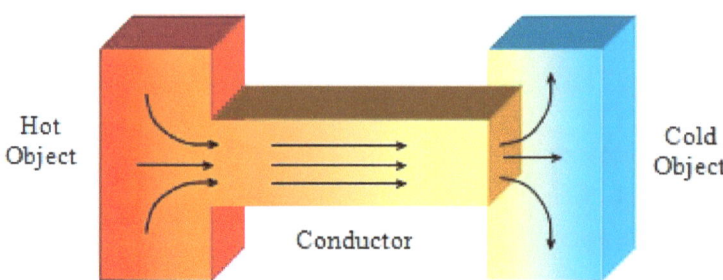

Figure 18.2. Just as heat flows from a hot to a cold object, energy flows from the Sun's core to a Stellar Core (SC) or from the earth's core to a SC, causing the energy fields of both objects moving toward the approaching energy depleted SC.

But photons can also be absorbed by matter, which shows that photons can exist as stationary particles; thus, light can become a part of matter. But it turns out that there is more to it than that because SCs create matter with energy, i.e. light (see Article 695: Planet X creates water: it started at the Flood) [2]. In other words, SCs and the core systems, inside all planets and stars, transform light into matter, which shows that matter is made of light. Thus, the whole universe is made of light, whether we think of the universe, and its processes, in terms of energy or in terms of matter, it is all made of light. Atoms are made of fluid, a malleable material, or plasma, i.e. a photon fluid.

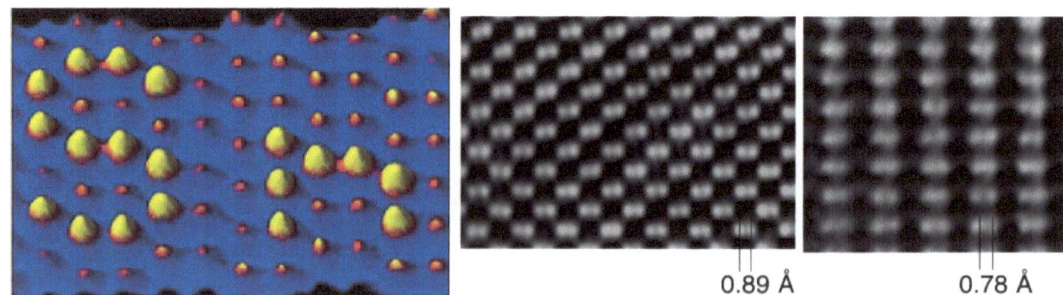

Figure 18.3. Electron microscope images: Atoms seem to have a regular shape, but they are not solid, they are made of light and can flow, and are thus, made of liquid plasma. **Left:** Atoms look like blobs of plasma. **Right:** the image on the far right is at a higher magnification and we can see a swirling flow of material connecting atoms indicative of a liquid plasma or photon fluid.

Energy transfers across a transparent medium as a photon fluid current and within matter as a photon fluid current as well, but within matter this transfer can manifest as heat flow or an electric current. The electric current manifests at the point at which the drawing object is located. In other words, a meteor draws a current from the earth's core but it only manifests at its location, not in between (see Article 779: Meteors are Planet X debris: glow is due to electric current) [3].

Figure 18.4. A meteor streaks across the sky on April 16th 2019.

In the same way a resistor, in a circuit, draws the current at its location, only, this is why current is not drawn by an object that has zero electric resistance, i.e. an insulator. This is also why when you flick a light switch the light goes on immediately, the energy (light) transferred through the circuit at the speed of light and the current was drawn at the light bulb location almost immediately, there were no electrons moving through the circuit at all, only photons, which take the form of electrons at the light bulb location only, in order to transfer energy to the light bulb. Energy is transferred to the light bulb because it has capacity to absorb energy, which is described as resistance. But energy is transferred to the light bulb because it has a capacity to draw energy and thus the word 'resistance' describes an opposing concept. Energy flows to the light bulb because it draws energy rather than resists. This is simply another example of how the educational system has been formulated to impede understanding of how the universe really works.

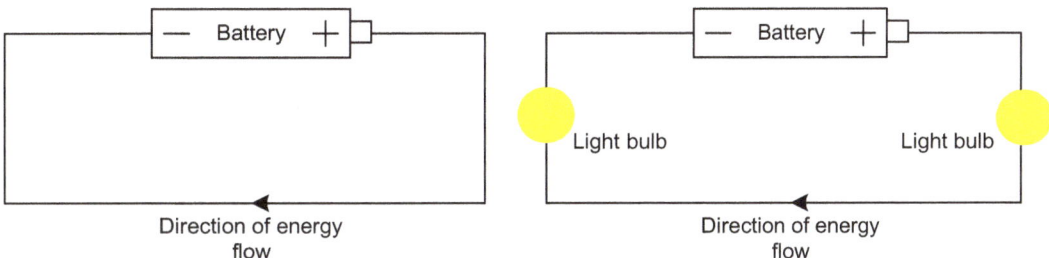

Figure 18.5. Left: Energy flows from one side of the battery to the other, and thus it will gradually discharge if a wire is connected from one side of it to the other. Even if no wire is connected the battery gradually discharges because its energy, light, flows through the air, but at a slower rate, the lower density of the air provides less pathways for the energy to travel through most likely. **Right:** If light bulbs are placed within the circuit the energy flowing across the wire, in the form of light, manifests as an electric current at the location of the light bulbs according to their capacity to absorb.

Meteors, inside the earth's atmosphere, work in exactly the same way; they draw an electric current at their location, within the atmosphere, because there is energy continuously transferring across all the matter that makes up the earth, including its atmosphere. Comets also draw a current whenever they come into the solar system, at their location, within the Sun's heliosphere (the most tenuous of the Sun's atmosphere going out to 100 au) because there is energy, or photons, continuously transferring across all the matter that makes up the Solar System.

Figure 18.6. A comet inside the Solar System draws a huge current from the Sun's core at its location in the Solar System due to energy, i.e. photons, continuously flowing across the Sun's tenuous atmosphere throughout the Solar System. The appearance of such an electric current indicates that the universe is electrical in nature but since this occurs because photons are transformed into electrons in these instances, the universe is still a light universe.

In conclusion, the whole universe is made of light or photons and all processes are due to the transfer of photons which we call energy. Comets and meteors glow and have tails because they have a capacity to absorb energy, i.e. they are made of energy depleted matter, and thus a part of the Planet X Systems coming into the Solar System. The energy, which Planet X matter draws, manifests as an electric current, at their location within the Solar System.

References:

[1]   Albers, C. (2019). Article 785: Planet X is here but what is it exactly?
[2]   Albers, C. (2019). Article 695: Planet X creates water: it started at the Flood.
[3]   Albers, C. (2019). Article 779: Meteors are Planet X debris: glow is due to electric current.

Chapter 19

779. Meteors are Planet X debris: glow is due to electric current

Figure 1 below shows some screenshots from video footage of a meteor in the sky. This meteor was seen from New York, New Jersey and the mid-Atlantic, on Tuesday night, April 16th 2019. The meteor becomes extremely bright, at one time, to the point that the sky lights up, to a bright blue color.

Figure 19.1. The meteor glows bright blue as it streaks across the sky. It then becomes even brighter, after which times, it starts to fade and eventually is seen as only a tiny white spot in the sky. This takes 4 seconds to occur suggesting that the meteor is moving at high speed as it follows a downward path toward the surface of the earth.

Now, according to the accepted explanation, meteors glow so brightly because they burn up in the atmosphere, due to friction or drag. However, friction can only cause two solid surfaces rubbing against each other to increase their temperature, not a surface within earth's gaseous atmosphere.

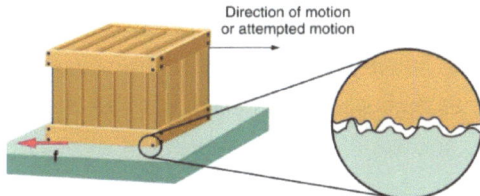

Figure 19.2. Friction occurs when two rough surfaces rub against each other, which can cause both to experience a temperature increase. A smooth surface will cause an object to slide across it, offering no resistance to its movement. But normal surface are rough and thus heat up as result of rubbing against

each other, because the molecules on the surfaces are like springs, and all molecules are strongly connected to each other so that all are like springs vibrating at an increased rate.

Temperature is defined as the degree of movement or vibration of molecules, thus the rough surfaces experience a temperature increase indicating that energy got transferred to both materials, which we can describe as it heating up.

But a solid object, moving through the atmosphere, cannot heat up, because the atmosphere is gaseous, not a solid surface, and thus the force that the gas molecules can exert cannot in any way deform the molecular structure of the solid surface like contact with another solid surface can.

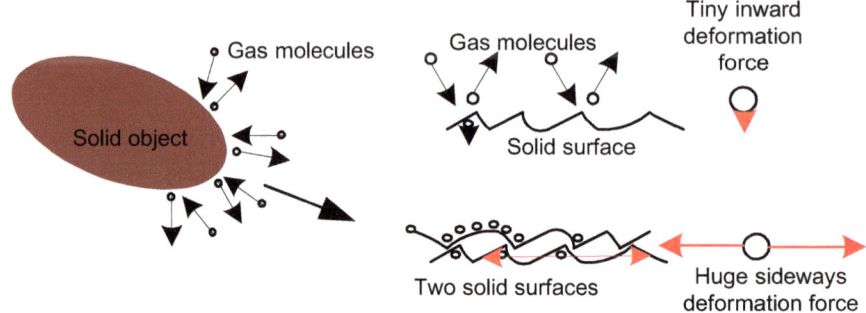

Figure 19.3. Gas molecules collide with the solid surface and bounce off. If the object is moving at high speed there will be more collisions, but they still cannot deform the surface and thus cause heating like two solid surfaces moving past each other, can.

If air friction could cause a solid object, moving through it, to heat up, then parachutes, which experience an extreme amount of drag or friction, when deployed in the earth's atmosphere, would heat up and go up in flames. Also an exposed face, on a fast moving motorcycle, does not heat up because of air being forced to part, as the face moves through it. Thus, meteors cannot burn up due to friction.

Meteors are Planet X debris, and they were a part of the inside layers of the planet. They were magma created by the core system of a Planet X planet, which solidified when the planet lost its energy supply and turned into solid rock, which then broke into pieces, as the planet broke up, due to the core system having become depleted in gravitational energy. Because they were deep inside the planet, they had a large amount of gravitational energy, which they lost at that time, which has left them with a high capacity for absorbing energy, most likely because they have a higher density than matter that was closer to the surface of the planet. This means that when meteors enter the earth's atmosphere, they are detected by the earth's core, as high density matter, with extreme energy depletion, for its density, in the outer layers of the planet, and thus a huge amount of energy flows towards the objects, in the form of an electron current, in order to re-establish equilibrium. The fast absorption of energy causes chemical reactions, on the object's outer surface layers, which in turn, causes the object to glow brightly until it has absorbed the gravitational energy, its density dictates, it should have.

Figure 19.4. Various photographs showing pieces of rock which fell from the sky over Cuba on February 1st 2019. The rocks are black on the outside due to the chemical reactions, which resulted from a larger electron current, flowing into it, in order to equalize its gravitational potential according to its density (see Article 595: Meteorite over Cuba: part of the Planet X debris field)[1].

Planet X debris, which had been on the surface of the planets, absorb energy at a much slower pace and thus sink through the atmosphere, at a much slower pace, as demonstrated by the pieces of rock with buildings on them, and icebergs, which remain suspended in the atmosphere, for long enough to be observed as cities in the sky (see Article 767: Planet X debris coming in as rocks, snow, hail and buildings) [2].

Figure 19.5. Left: A huge rock appears amongst the clouds in Peru: This piece of rock is long and flat and obviously large enough to have had many buildings upon it, had it been a part of the planet, to which it once belonged that had had buildings on it? The cities seen in the sky do not seem to be built on thick pieces of rock, and the fact that this is a long flat piece, suggests that the surface of these planets may have broken into pieces of rock that were long and flat. **Right:** Buildings within clouds are seen floating in the sky, over the ocean: The buildings are seen over a region where there is nothing but ocean. These buildings are on a large piece of rock from the surface of one of the Planet X planets, which were destroyed due to gravitational energy depletion (see Article 584: Planet X the reason behind GMO crop) [3].

All Planet X surface material, including ocean water, which froze into blocks of ice, when the planets were destroyed, will also slowly absorb energy, once inside the earth's atmosphere, and will thus slowly sink down through the atmosphere, until it has reached the surface. This frozen ocean is what is referred to as icebergs, once they enter the ocean. They are also, most likely, being steered to Antarctica by the 'powers that be' so that they end up as close to the South Pole, and thus as far away from prying eyes, as possible, and are thus constantly replenishing the Antarctica ice sheet (see Article 777: Icebergs are Planet X frozen ocean: falling in the ocean from the sky and Article 773: Planet X: the source of earth's polar ice caps) [4, 5].

Figure 19.6. An iceberg floats in the ocean. Icebergs are Planet X frozen ocean and come in from space [4].

In conclusion, meteors glow brightly inside the earth's atmosphere, because they are high density, energy depleted matter, coming to the earth as part of the Planet X debris field, which attract a strong energy flow, in the form of an electric current, once inside the earth's atmosphere. The illogical explanation about meteors burning up due to air friction is another attempt by the 'powers that be' to cover up the truth about Planet X.

References:

[1]   Albers, C. (2019). Article 595: Meteorite over Cuba: part of the Planet X debris field.
[2]   Albers, C. (2019). Article 767: Planet X debris coming in as rocks, snow, hail and buildings.
[3]   Albers, C. (2019). Article 584: Planet X the reason behind GMO crops.
[4]   Albers, C. (2019). Article 777: Icebergs are Planet X frozen ocean: falling in the ocean from the sky.
[5]   Albers, C. (2019). Article 773: Planet X: the source of earth's polar ice caps.

Chapter 20

1166. Gravitational potential and the size of earth's central core

Isaac Newton came up with the Universal Law of Gravitation, which is based on Kepler's laws. Kepler's laws are based on observations of the motion of the planets in the Solar System and thus strictly mathematical, i.e. they do not pretend to explain what causes the gravitational interaction, but simply describe the motion that gravity causes in an interacting system, such as the Solar System. So, this seems to be a good point to start in attempting to describe the gravitational interaction. According to Newton's Universal Law, gravitational potential energy is given by

$$U_G = -\frac{GMm}{r} \quad (1)$$

where M and m are the masses of two interacting objects, r is the distance between them and G is the gravitational constant, which Planet X observations have shown is not really a constant but dependent on energy. In addition, due to gravity's close alliance with electricity, I am going to define an electrical potential, using a similar scheme to what is used in electrostatics, i.e.

$$V_G \equiv -\frac{U_G}{m} = \frac{GM}{r} \quad (2)$$

which describes the gravitational potential of an object of mass M, at a point, in space, which is a distance r from its center. In this scheme, gravitational potential is equal to the orbital velocity squared, of an object orbiting M, at an orbital distance r. We can see that by applying Newton's second law to an object orbiting M, in a circular orbit, at a distance r. Since the magnitude of the gravitational force is given by

$$F_G = \frac{GMm}{r^2} \quad (3)$$

Then, applying Newton's second law, i.e. applying: $\vec{F}_R = m\vec{a}_R$ we get, along the radial direction:

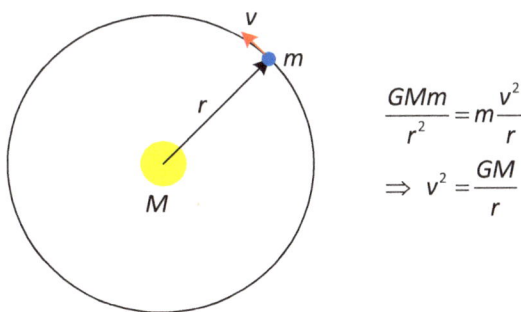

where *v* is the tangential speed of the orbiting object, m, i.e. the magnitude of its orbital velocity.

So, gravitational potential will have units of velocity squared. Using the definition from equation (2), the earth's gravitational potential, in terms of its orbit around the Sun, is given by

$$V_{GE} = \frac{GM_S}{r_{ES}} \qquad (4)$$

where M_S is the Sun's mass and r_{ES} is the earth's orbital radius. And, the earth's gravitational potential in terms of its own gravitational field is given by:

$$V_{GE} = \frac{GM_E}{r_{EC}} \qquad (5)$$

where M_E is the earth's mass and r_{EC} is the radius of the earth's central core, where the analogy to what happens to the electric potential, in the case of a conducting sphere, which is constant inside the sphere and equal to the value of the potential on its surface, has been used.

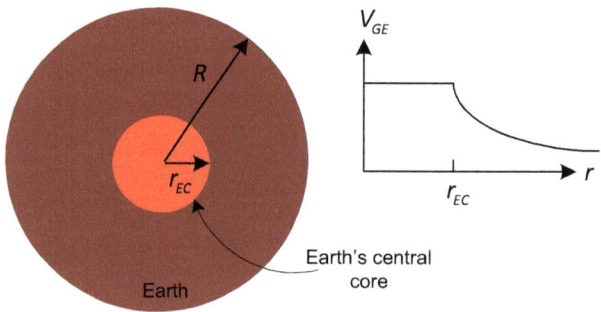

Figure 20.1. In analogy to what happens with electrical potential in the case of a spherical conductor whose surface electrical potential is constant on the inside and equal to its surface electrical potential, we set the earth's maximum gravitational potential to be the gravitational potential on the surface of its central core, which is thus seen as a perfect gravitational conductor. If we do not make the potential constant inside the central core, we would get an infinite potential at the center of the earth, which cannot be correct.

Since we now have two equations for earth's gravitational potential, we can use them to find the radius of earth's central core:

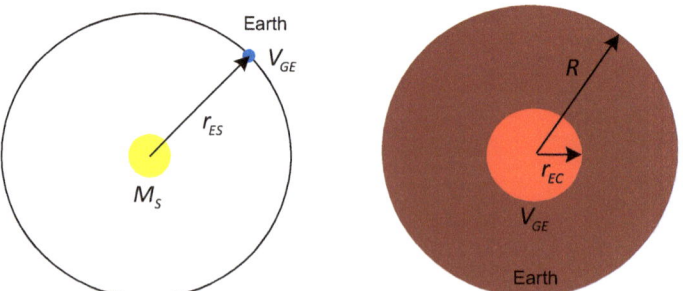

Figure 20.2. We have two ways to obtain earth's gravitational potential.

$$V_{GE} = \frac{GM_S}{r_{ES}} = \frac{GM_E}{r_{EC}} \Rightarrow r_{EC} = \frac{M_E}{M_S} r_{ES} = 450 \text{ km}$$

Or 279 miles, where the following have been used: M_E = 5.972 x 10^{24} kg, M_S = 1.989 x 10^{30} kg and r_{ES} = 150 x 10^6 km. Now, 450 km is quite a bit smaller than the accepted value (1200 km) and the value that I have estimated in the past. However, this is not a result that is anywhere near to incredible. Cores of about this size were observed inside Mercury's orbit, one of these had a radius of 391 km (243 miles) and the other of 938 km (582 km) (see Article 1028: Earth's new moons and size of Planet X objects observed in 1800s) [1]. If these are typical planetary central core sizes, then we can see that the earth's central core of the size obtained above, would be somewhere inside that range, and a little towards the lower end, i.e. it would indicate that the earth was initially a small planet, as planets go in the universe.

Now, I had previously stated that the earth's central core must be larger than the moon (see Article 1106: Moon's tidal pattern and the size of Earth's central core: 1745 km) [2] but from this result it seems that this is not so. However, the planet's solid body seems to be an extension of the central core, i.e. it is like an outer envelope, but still a part of the whole, just like each trovant seems to have an inner structure and an outer layer.

Figure 20.3. Trovants: rocks that grow and reproduce because they are cores. Cores are made of the same matter as created matter but they are able to use energy to create more matter. This suggests that cores have a spiritual component which normal matter does not possess.

It is thus likely that in terms of the gravitational connections, which occur between Planet X cores and the earth's core system that it is the overall size of the earth that counts. This means that as far as the moon (moon's radius is 1731 km and the earth's overall radius is 6371 km) is concerned, it is connecting to a much larger core than it is; which then results in a gravitational tidal connection with a central minimum [2].

In conclusion, in an attempt to use what we already know about gravity, which has not been refuted by the observation of Planet X cores and the discovery that all celestial objects have core systems, I start with Newton's Law of Universal Gravitation and in analogy to electrical potential, as it is used in electrostatics, I define a gravitational potential and use it to determine the size of the earth's central core.

References:

[1] Albers, C. (2019). Article 1028: Earth's new moons and size of Planet X objects observed in 1800s.

[2] Albers, C. (2019). Article 1106: Moon's tidal pattern and the size of Earth's central core: 1745 km.

Chapter 21

1167. How could astronomers have missed this? Gas giants tell us G is not constant

The known gas giants in the Solar System are Jupiter, Saturn, Uranus and Neptune, and they are called gas giants because they are believed to be made mainly of gas. Astronomers have had these objects' orbital parameters for a very long time. Some of these orbital parameters are shown in table 1 below, namely the radius and average orbital radius of each gas giant.

Figure 21.1. The four gas giants in the Solar System: Jupiter, Saturn, Uranus and Neptune.

Table 21.1. Orbital parameters for the four gas giants: radius, in terms of the Sun's radius, and orbital radius in astronomical units (1 au = distance between the earth and the Sun).

Gas Giant	$R\ (R_{Sun})$	r (au)
Jupiter	0.1	5.2
Saturn	0.0837	9
Uranus	0.0365	19.2
Neptune	0.0354	30.1

Now, when the radius, R, of each gas giant is plotted against the orbital radius, r, a very clear relationship is found, which is given by:

$$R \propto r^{-2/3} \tag{1}$$

The way this relationship is found is by assuming the relationship: $R = r^x$, where x is unknown. We then take natural logs on both sides to obtain: $\ln R = x \ln r$ and plot $\ln R$ versus $\ln r$, the slope of the graph then gives us the value of x. When this is done, the value obtained for x is − 0.675, which only differs by 1.2%, from -2/3 = - 0.666666.., which indicates that this is the true relationship. That the measured orbital radius and radii of these 4 planets have such an unmistakable relationship cannot be a mistake. So why haven't the astronomers who have had this data for so long not found it? How could they have missed such a simple relationship?

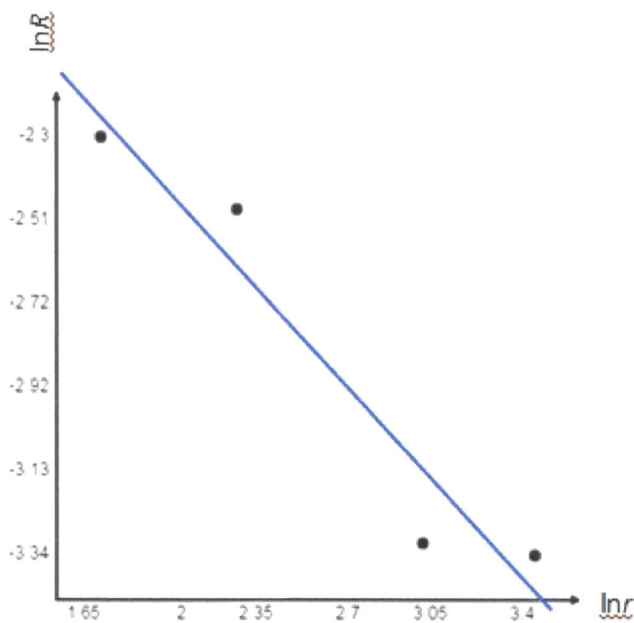

Figure 21.2. The slope to the graph is -0.675, which is extremely close to -2/3, so we take that to be the relationship between the radius and the orbital radius of the gas giants. Then, $R \propto r^{-2/3} \Rightarrow R^3 = kr^{-2}$, i.e. the volume ($\frac{4}{3}\pi R^3$) is inversely proportional to the orbital radius squared.

Now, the fact that radius R is dependent on the orbital distance r, means that the size of each gas giant is determined by its distance from the sun. This is only possible, if they all have approximately the same density, which tells us that they cannot be made of gas. They have to be solid objects with a thin gaseous atmosphere. And, this agrees with the observation that the Shoemaker-Levy 9 comet impact, in July of 1994, left very long lasting blemishes on Jupiter, which would have been impossible had there not been a solid surface, right underneath the top of Jupiter's gaseous atmosphere. This was understood by Dr. Eugene Shoemaker and I believe that was the reason why he was killed (see Article 621: Gas Giant vindication and why Dr. Eugene Shoemaker was really killed) [1].

Since they have to be solid objects, the gas giants, due to their large size, should have densities even higher than the rocky planets, in the Solar System. Earth's average density is 5.51 g/cm³, which is much higher than the densities of all the gas giants, shown below.

Table 21.2. Orbital parameters for the four gas giants: radius, orbital radius and density.

Gas Giant	R (R_{Sun})	r (au)	P (g/cm³)
Jupiter	0.1	5.2	1.33
Saturn	0.0837	9	0.687
Uranus	0.0365	19.2	1.27
Neptune	0.0354	30.1	1.64

So there has to be something wrong with these calculated densities. By looking at the equation used to calculate the density, we can see what the real problem is. The density of each of the gas giants is calculated from the period and orbital radius of its satellites, using Kepler's third law:

$$T^2 = \frac{4\pi^2}{GM}a^3 = \frac{4\pi^2}{G\rho\frac{4}{3}\pi R^3}a^3 = \frac{3\pi a^3}{G\rho R^3}$$

$$\rho = \frac{3\pi a^3}{GR^3 T^2}$$

(2)

where a is the satellite's orbital radius, T is its orbital period, R is the gas giant's radius and G is the gravitational constant. Now, a, R and T are obtained from observational measurements and cannot be argued with. So there is only one thing left that can lead to the obtaining of false densities and that is the gravitational constant. Clearly, G cannot be constant and has to be much lower for the gas giants, which agrees with what I have been stating for a very long time. The gas giants are Planet X central cores, Planet X star central cores, in fact. Planet X stars and planets were destroyed because of energy depletion and the strength of the gravitational interaction, which is determined by G, is dependent on energy. These objects even after having been adopted by the Sun, as if they are a part of the Sun's core system, remain energy depleted and thus their gravitational constants are lower, than what they are for original Solar System objects (see Article 1030: Mercury is not a planet: it is a part of the Planet X System) [2].

Since Jupiter is essentially a core, a star core, we can assume that it would at least have the same density as earth's core, which is estimated to have a density of 12.8 g/cm³, and we can then use that value to estimate what G is for Jupiter. Since $G\rho$, in equation (2) must be equal to a constant, we have:

$$G_J \rho_J = G\rho_m \Rightarrow G_J = \frac{\rho_m}{\rho_J}G = \frac{1.33}{12.8}G = 0.1G$$

where G_J is Jupiter's true gravitational constant, ρ_J is Jupiter's density which is set to be equal to that of earth's core, ρ_m is Jupiter's measured density, and G is the known gravitational constant for the original Solar System objects.

Table 21.3. Orbital parameters for the four gas giants: radius, orbital radius, density and gravitational constant (strength of gravitational interaction).

Gas Giant	R (R_{Sun})	r (au)	P_m (g/cm³)	G
Jupiter	0.1	5.2	1.33	0.10
Saturn	0.0837	9	0.687	0.054
Uranus	0.0365	19.2	1.27	0.099
Neptune	0.0354	30.1	1.64	0.12

Thus, Saturn has the lowest G and is therefore the most energy depleted of all the gas giants, which are not made of gas, at all; they are all Planet X star central cores, which have been adopted by the Sun.

In conclusion, there is a clear and unmistakable mathematical relationship between the radius and the orbital radius of the four known gas giants in the Solar System, which results in the volume of each gas giant being inversely proportional to their orbital radius. The only way that astronomers could possibly have missed such a clear relationship is if they have ignored it on purpose. This relationship leads to the conclusion that gas giants cannot be made of mainly gas and that the gravitational constant, G, is not constant.

References:

[1]   Albers, C. (2019). Article 621: Gas Giant vindication and why Dr. Eugene Shoemaker was really killed.

[2]   Albers, C. (2019). Article 1030: Mercury is not a planet: it is a part of the Planet X System.

Chapter 22

1171. How did the Planet X planets break into pieces?

The Planet X planets broke into pieces due to energy depletion. All celestial objects have a core system in them and this core system has a central core and satellite cores, which create all the matter that the planet is made of. The inside part of the planet broke into small pieces and the outer layers broke into large pieces. We can see this from the fact that large flat pieces come in with buildings on them, indicating that these are surface pieces, which also absorb energy slowly and thus sink through the earth's atmosphere slowly. Smaller pieces of the broken planets come in as meteors and fall through the atmosphere at a fast pace (see Article 1164: Interstellar comet: Planet X star system coming in) [1].

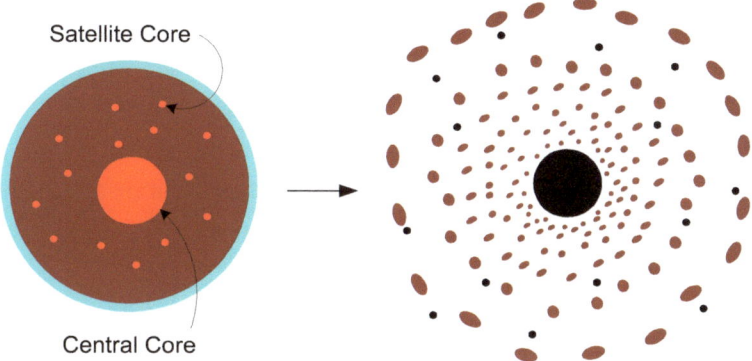

Figure 22.1. The Planet X planets broke apart due to energy depletion, the inner layers broke into smaller pieces and the surface pieces into larger pieces.

Now, the pieces of broken planet and the cores did not experience a drop in density but the planet as a whole did and the volume occupied by the planet increased. So, if the planet's average density, before breaking up, was ρ_o, and if its volume, before breaking up, was V_o, the density and volume after breaking up could be described by:

$$\rho = \frac{1}{N}\rho_o \quad \text{and} \quad V = NV_o \tag{1}$$

where N is a large positive number. Then, the mass of the planet would be given by

$$m = \rho V = \frac{1}{N}\rho_o NV_o = \rho_o V_o = m_o \tag{2}$$

Thus, the mass of the planet would not change, and in addition, the densities of each piece of rock or soil remained the same. However, each piece's gravitational potential had to have dropped in order for their position, with respect to the central core, to have increased. Since the volume changed as shown by equation (1) and since $V = \frac{4}{3}\pi r^3$ we have that

$$V = NV_o \implies \tfrac{4}{3}\pi r^3 = N\tfrac{4}{3}\pi r_o^3 \implies r = N^{1/3} r_o \tag{3}$$

which gives the distance, by which each piece of planet, moved away from its original position. Now, we cannot know for sure by how much the gravitational potential of the planet dropped, over all, but let us suppose it dropped by a factor of *N*, like the density, then

$$V_G = \frac{1}{N} V_{oG} \tag{4}$$

Now, initial and final gravitational potentials will be given by (see Article 1166: Gravitational potential and the size of earth's central core) [2]:

$$V_G = \frac{Gm}{r} \quad \text{and} \quad V_{oG} = \frac{G_o m_o}{r_o} \tag{5}$$

Thus,

$$V_G = \frac{1}{N} V_{oG} \Rightarrow \frac{Gm_o}{N^{1/3} r_o} = \frac{1}{N} \frac{G_o m_o}{r_o} \Rightarrow G = \frac{G_o}{N^{2/3}} \tag{6}$$

which shows that the planet's gravitational constant would decrease due to the energy depletion (see Article 1167: How could astronomers have missed this? Gas giants tell us G is not constant) [3]. Thus, G is not really a constant, it is the strength of the gravitational interaction, and it is dependent on energy. But, in addition, Planet X cores can only approach Solar System objects to a minimum distance, which shows that there is a repelling gravitational force that becomes stronger than the attractive gravitational force at a certain minimum distance. This means that the gravitational 'constant' can be described in the following way:

$$G = G_a(E) - G_r(E, d_{min}) \tag{7}$$

where G_a gives rise to the attractive part of the gravitational interaction and is dependent on energy, *E*, and G_r gives rise to the repulsive part of the gravitational interaction and is dependent on both energy and a minimum distance, d_{min}, of closest approach to a neighboring celestial object or another core. Thus, the repulsive term is equal to zero unless the distance of closest approach is reached, at this point the repulsive term becomes infinite, so that the two objects cannot approach beyond that point.

Now, when the planet broke apart due to a drop in gravitational potential, the attractive part of the interaction, or G, would have decreased, and at the same time the distance of close approach, for each core in the system increased, causing them to move apart and all the matter they had created to break into pieces. The repulsive force must have overtaken the attractive, even faster in the inner layer, which then caused it to break into smaller pieces.

Now, equation (7) gives the appropriate function for G, which describes the gravitational potential due to one core, or one celestial object. But, often we need to describe the interaction between two objects, such as when we are looking for the magnitude of the gravitational force, between two masses, which is given by:

$$F_G = \frac{GM_1 M_2}{r^2} \tag{8}$$

In this case, G will be given by:

$$G = \left[G_{1a}(E_1) - G_{1r}(E_1, d_{1min})\right]^{1/2} \left[G_{2a}(E_2) - G_{2r}(E_2, d_{2min})\right]^{1/2} \tag{9}$$

which describes the interaction in the case of the interacting masses having a core system or being core. If the interacting masses are created matter than there would not be a repulsive term in the expression, as created matter does not generate a repulsive gravitational interaction.

In conclusion, it has now become possible to mathematically describe the break-up of the Planet X planets, and how such an event would lead to a drop in the gravitational constant, which is really a measure of the strength of the gravitational interaction.

References:

[1]   Albers, C. (2019). Article 1164: Interstellar comet: Planet X star system coming in.
[2]   Albers, C. (2019). Article 1166: Gravitational potential and the size of earth's central core.
[3]   Albers, C. (2019). Article 1167: How could astronomers have missed this? Gas giants tell us G is not constant.

Chapter 23

1173. Determining Venus' true mass with the help of Planet X

The mass of a planet is determined from the orbit of one of its satellites or moons but Venus does not have a satellite so computing its mass is not so easy. Venus' mass was calculated using perturbations to the orbits of asteroids. One paper detailing the results of one of these determinations is shown below. There is however a big problem with using asteroids because asteroids are Planet X debris and since the Planet X planets were destroyed as a result of energy depletion, all the Planet X debris and thus all asteroids, in the Solar System, are energy depleted, which causes the gravitational constant, or strength of the interaction between asteroids and Solar System planets to be weaker than if they had not been energy depleted (see Article 1141: Asteroids coming in due to Planet X: Earth being destroyed) [1]. Thus, the use of an asteroid is not likely to yield the correct mass for Venus, the mass will be lower than the true value.

Determination of Masses of Mercury and Venus from Observations of Five Minor Planets

by

G. Sitarski

Space Research Center, Polish Academy of Sciences, ul. Bartycka 18a, 00-716 Warsaw, Poland
e-mail: sitarski@cbk.waw.pl

Received June 8, 1995

ABSTRACT

We collected 1985 astrometric observations of the five minor planets: (1566) Icarus, (1620) Geographos, (1862) Apollo, (2101) Adonis and (2212) Hephaistos. The asteroids can closely approach Mercury and Venus. The observations cover the time interval from 1930 for Apollo up to 1995 for Adonis. Equations of motion of the asteroids have been integrated by the recurrent power series including all the planetary perturbations, perturbations caused by the four biggest minor planets and relativistic effects. Joining all the observational equations for the five asteroids we created a set of thirty two normal equations to determine by the least squares method thirty corrections to orbital elements of the five asteroids along with two corrections to the planetary masses. Thus we found the following values of the reciprocal masses for Mercury and Venus in the solar unit mass:

$$m_M^{-1} = 6019522 \pm 9964, \qquad m_V^{-1} = 408522.85 \pm 8.80.$$

Key words: *Planets and satelites: individual: Mercury, Venus – Minor planets, asteroids*

Figure 23.1. Screenshot of top of the first page of a paper detailing the calculation for Mercury's and Venus' mass based on perturbations to the orbit of asteroids [2]. Mercury's mass would be an even worse problem than Venus, as Mercury also has no natural satellites but it is also not a native planet to the Solar System. Mercury is a Planet X planetary central core (see Article 1030: Mercury is not a planet: it is a part of the Planet X System) [3]. Thus, in the case of Mercury, we would need to find its G, as well as its mass. The mass given above is $m_V M_{Sun} = 4.869 \times 10^{24}$ kg where $M_{Sun} = 1.989 \times 10^{30}$ kg.

Now, Venus' mass could have been determined from the orbit of probes sent to orbit it. However, since NASA is the frontline of the cover up for Planet X, it is unlikely that real data regarding a probe's orbit would become available that would yield Venus' true mass. This is to be expected as humanity's knowledge base is full of lies in order to cover up the truth about gravity and Planet X (see Article 1040: Aliens have filled humanity's knowledge base with lies) [4]. The current accepted mass for Venus, supposedly obtained from the orbit of Venus probes is 4.867 x 10^{24} kg, which is actually a little lower than the value obtained in the above paper.

Now, in Article 1166: Gravitational potential and the size of earth's central core [5], I defined gravitational potential as

$$V_G \equiv \frac{GM}{r}$$

where M is the mass of the object generating a gravitational field, r is the radial distance from its center and G is the gravitational constant for that object, and then used it to calculate the size of earth's central core, which turned out to be 450 km (279 miles) in radius. In this article, I am going repeat the calculation for the radius of the central core, for Mars, Pluto and Venus.

Table 23.1. Mass and orbital radius for the original (native) Solar System planets to be used to determine the radius of each planet's central core. The mass of Venus is however not expected to be correct.

Gas Giant	M (x 10^{24} kg)	r (x 10^6 km)	r (au)
Venus	4.867	108.2	0.723
Earth	5.972	149.6	1.00
Mars	0.639	227.9	1.52
Pluto	0.0131	5906	39.5

In order to calculate the size of each planet's central core, the gravitational potential of each planet is determined based on its position in the Sun's gravitational field as well as on its own gravitational field. These two are then equated. This is illustrated below for the case of earth.

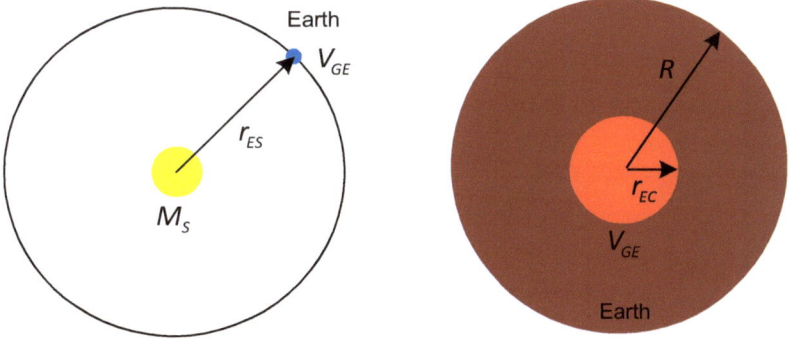

Figure 23.2. We have two ways to obtain earth's gravitational potential.

$$V_{GP} = \frac{GM_S}{r} = \frac{GM}{r_{PC}} \Rightarrow r_{PC} = \frac{M}{M_S} r$$

where r_{PC} is the planet's central core radius. Thus, the calculation in the case of Mars yields:

$$r_{PC} = \frac{6.39 \times 10^{23}}{1.989 \times 10^{30}} 227 \times 10^6 \text{ km} = 73.2 \text{ km} \quad (45 \text{ miles})$$

This makes sense as Mars is further away from the Sun than earth and would thus be expected to have a smaller central core to go along with its lower gravitational potential.

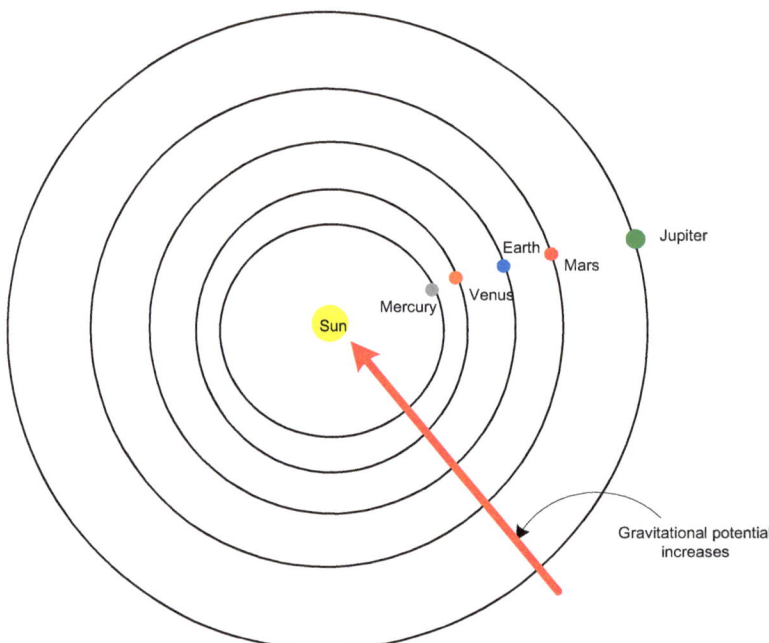

Figure 23.3. Planets orbiting further out from the Sun must have a lower gravitational potential. Jupiter and Mercury are not original Solar System objects, Mercury is a huge Planet X planetary central core and Jupiter is a Planet X star central core and so are the other so called gas giants. They are not made of gas (see Article 1167: How could astronomers have missed this? Gas giants tell us G is not constant) [6].

Table 2 below shows the result of repeating the above calculation for the other planets:

Table 23.2. Calculated central core radii (R) for the original Solar System planets using the values shown for mass and orbital radius. Venus' central core radius is lower than Earth's, which cannot be correct.

Gas Giant	M (x 10^{24} kg)	r (x 10^6 km)	r (au)	R = r_{PC}
Venus	4.867	108.2	0.723	265
Earth	5.972	149.6	1.00	450
Mars	0.639	227.9	1.52	73.2
Pluto	0.0131	5906	39.5	39

We see from this result that Venus' mass has to be wrong. So, in order to determine the correct central core radius and mass for Venus, we will plot the central core radius versus orbital radius, for the other 3 planets: Earth, Mars and Pluto, to see what the relationship is. So, assuming the relationship: $R = kr^x$; then taking natural logs on both sides we get: $\ln R = x\ln r + \ln k$ which is the equation of a straight line. When the data is plotted, the slope of the graph gives us x and the y-intercept gives us k.

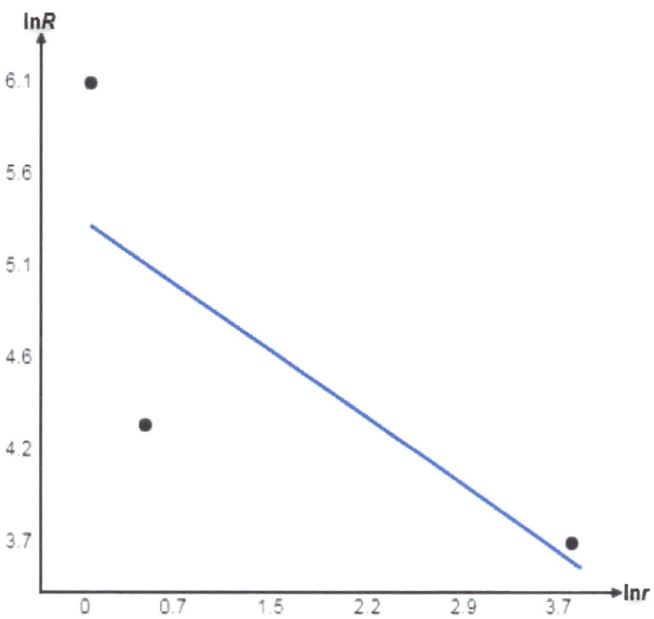

Figure 23.4. It is not ideal to do this kind of thing when there are only 3 data points available but one uses what is available. The slope for the line that best fits the 3 data points is - 0.486 (or - 0.5) and the y intercept for it is 5.35. Thus x = - 0.5 and k = 211.

Hence, the relationship between R and r is given by

$R \propto r^{-1/2}$

This is a different relationship to what was found with the gas giants, which was [5]

$R \propto r^{-2/3}$

This indicates that the orbital position for the gas giants (Planet X star cores) will increase faster as the size of their cores decreases, i.e. they will be much further from the Sun then planetary cores of the same size. Since the gas giants are star cores, this suggests that star-star gravitational interactions and star-planet gravitational interactions are different. Stars will remain further from each other. This will be due to the repulsive part of the gravitational interaction being stronger for star-star interactions than for star-planet interactions and explains why the gas giants being basically just solid star cores are so far out in the Solar System, whilst Mercury, which is much smaller is so close to the Sun. Mercury is a planetary core, although one of an extremely large planet, as that particular core is 5.4 times larger than earth's central core. Now, since the earth expanded by 82% during the flood cataclysm, so that earth's original radius was estimated to be 3490 km (2170 miles), earth was originally 7.75 larger than its central core, so the planet that Mercury would have been a central core to, would have had a radius of 17 360 km (10

800 mi) and thus be 2.7 times larger than the earth currently is (see Article 1105: Earth's radius increased by 82% during the Flood cataclysm due to Planet X) [7].

Now, using the relationship between central core radius and orbital distance we can find the true size of Venus' central core:

$$\frac{R_V}{R_E} = \frac{r_V^{-1/2}}{r_E^{-1/2}} = \left(\frac{r_E}{r_V}\right)^{1/2} \Rightarrow R_V = \left(\frac{1}{0.723}\right)^{1/2} (450 \text{ km}) = 529 \text{ km } (328 \text{ miles})$$

Then, we can calculate Venus' mass from the gravitational potential equations:

$$V_{GV} = \frac{GM_S}{r_V} = \frac{GM_V}{R_V} \Rightarrow M_V = \frac{R_V}{r_V} M_S = 1.101 \times 10^{25} \text{ kg}$$

Table 3. Calculated central core radii (R) for the original Solar System planets and corrected values for Venus' central core radius as well as its mass.

Gas Giant	M (x 10^{24} kg)	r (x 10^6 km)	r (au)	R = r_{PC}
Venus	11.01	108.2	0.723	529
Earth	5.972	149.6	1.00	450
Mars	0.639	227.9	1.52	73.2
Pluto	0.0131	5906	39.5	39

Venus as a planet orbiting the sun at a much closer distance to the Sun than the earth would logically be expected to be more massive than the earth, as it must have a much higher gravitational potential. Venus is however smaller than the earth in terms of radius, which suggests that it did not go through the huge expansion that the earth did, at the time of the flood cataclysm, i.e. Venus is still the original size that it had when it first formed.

In conclusion, using the understanding of gravity and the Solar System gained through Planet X observations, it is now possible to obtain the size of the central cores of all the original Solar System planets, as well as an estimate of Venus' true mass.

References:

[1] Albers, C. (2019). Article 1141: Asteroids coming in due to Planet X: Earth being destroyed.

[2] Sitarski, G. (1995). Determination of Masses of Mercury and Venus from Observations of Five Minor Planets. Acta Astronomica, v.45, pp.665-672.

[3] Albers, C. (2019). Article 1030: Mercury is not a planet: it is a part of the Planet X System.

[4] Albers, C. (2019). Article 1040: Aliens have filled humanity's knowledge base with lies.

[5] Albers, C. (2019). Article 1166: Gravitational potential and the size of earth's central core.

[6] Albers, C. (2019). Article 1167: How could astronomers have missed this? Gas giants tell us G is not constant.

[7] Albers, C. (2019). Article 1105: Earth's radius increased by 82% during the Flood cataclysm due to Planet X.

Chapter 24

1180. The photon model: all isolated objects have an aura

As I detailed in Book 3: Planet X revealed Gravity and Light [1], Planet X observations lead to the understanding that the gravitational interaction comes from within light or photons and it causes all isolated agglomerations of photons, i.e. matter, to have an outer negatively charged layer, from atoms to galactic nuclei. So, planets and stars also have an outer negatively charged layer.

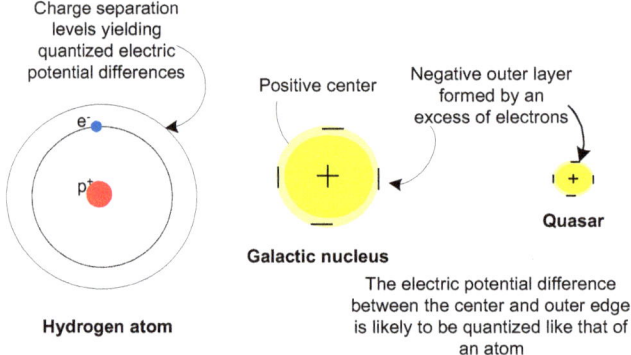

Figure 24.1. All isolated matter agglomerations have a negative outer layer; atoms have an outer layer of electrons. The electrons and the protons both emerge from within photons and are thus made of light.

Thus, it seems logical that at all levels, gravitational energy within matter causes it to split into 2 parts, a positive part surrounded by a negative part, so that planets become concentric spherical capacitors.

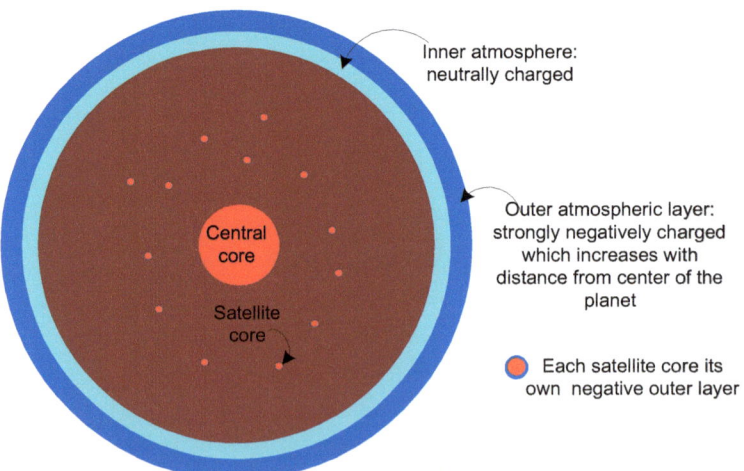

Figure 24.2. A planet will have an outer negative layer but so will each core. Thus, all cores will be surrounded in a negative electric field as will all agglomerations of photons, not moving at light speed,

i.e. all atoms and all molecules will be surrounded in an outer negative field, produced by electrons. The planet's outer negative layer is the central core's negative outer layer.

But, all isolated agglomerations will also be surrounded by a negative layer and thus have a negative electric field surrounding it, this means all objects and all living things will have a negative electric field surrounding them. Droplets of water since they are isolated, i.e. there is a space separating each droplet of water from other droplets, will also have a negative electric field around them. This means that all isolated matter will be in the form of a capacitor, including the human body; there will always be an electric potential difference between just inside the skin of the human body and just outside the skin. This is the source of the human aura.

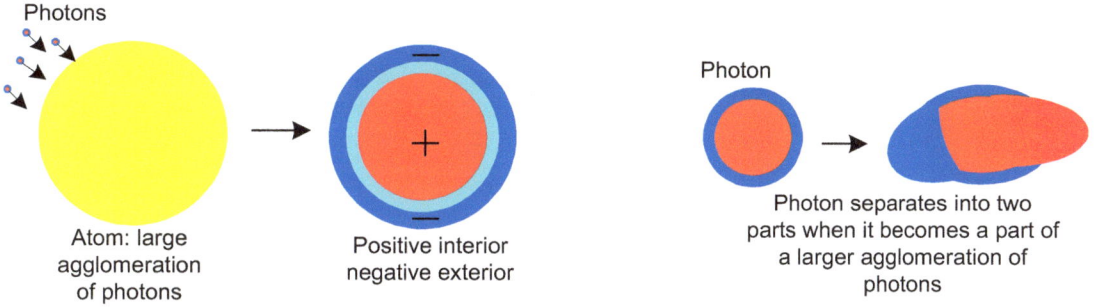

Figure 24.3. An atom is made of a large number of photons which can absorb photons. Once absorbed, photons increase the energy of an atom, and cause electrons to move to higher orbits, thus the electron layer increases in size and so does the nuclear matter, which suggests that we are dealing with a liquid plasma, not individual point like particles, the absorbed photon splits into 2 parts and adds to the plasma making up the nucleus, as well as the plasma making up the outer layer.

Since photons are electromagnetic waves and thus are not only carriers of the electric field, which arises from them being made up of a positive and a negative type of material; they must also have a magnetic property. This magnetic property most likely arises from the photon being toroidal or disk shaped and it spinning. Also, since photons must have oscillating magnetic and electric fields, they must oscillate with respect to which plasma is on the inside and which is on the outside.

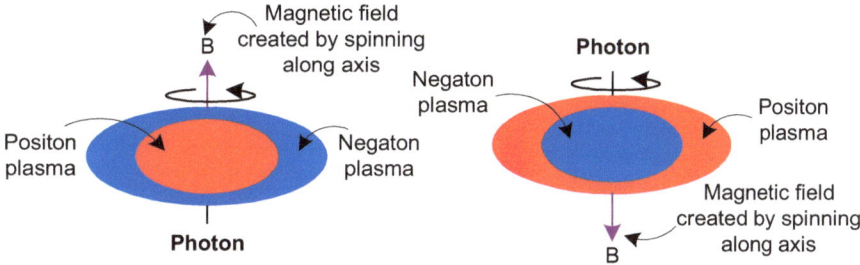

Figure 24.4. Photon model: Photon is made of two types of plasma, one is positively charged and the other negatively charged, which makes the photon overall neutral. Both the positive and the negative parts generate a magnetic field in opposite directions, due to the spin, but the one on the outside generates the largest magnetic field, which becomes the overall magnetic field of the photon. Photons fluctuate between the two different arrangements causing the magnetic field to fluctuate between the

upward and the downward directions as the electric field also fluctuates. The plasma is massless as long as the photon moves at the speed of light.

Once the photon is absorbed, i.e. not moving at the speed of light, anymore, its energy gets converted to mass and/or gravitational energy. The positon plasma gains more mass, about 2000 times more (as the proton is 1837 times more massive than the electron) than the negaton plasma, which causes the photon to adopt the positon on the inside configuration. It takes a huge number of single photons to create a single hydrogen atom, which is made up of one proton and one electron. To find out how many let us suppose we want to construct a hydrogen atom with green light photons of wavelength 550 nm. These photons have a frequency of 550 THz or 5.50 x 10^{14} Hz. So the energy of this photon is then given by

$$E_{ph} = hf = (6.63 \times 10^{-34})(5.50 \times 10^{14}) \text{ J} = 3.6 \times 10^{-19} \text{ J}$$

where h is Planck's constant and f is frequency. Then since the mass of a proton is 1.6726 x 10^{-27} kg, the energy associated to the mass of one hydrogen atom is

$$E_H = (m_p + m_e)c^2 = m_p \left(1 + \tfrac{1}{1837}\right)c^2 = 1.5 \times 10^{-10} \text{ J}$$

where the fact that the electron is 1837 times less massive than the photon has been used and c is the speed of light, which is given by 3 x 10^8 m/s. Then we can determine how many photons it would take to create one hydrogen atom from

$$\frac{E_H}{E_{ph}} = \frac{1.5 \times 10^{-10}}{3.6 \times 10^{-19}} = 400 \times 10^6$$

So, it takes 400 million visible light photons to create one hydrogen atom. This is a lot of photons.

Thus, Planet X observations lead to a matter model based on the photon, i.e. the universe is made of photons (light) (see Article 818: Planet X observations: the electrical universe is made of light) [2], which already contain within them everything necessary to give rise to all matter, energy and interactions we observe in the universe. This includes all the known interactions, namely the gravitational, electric and magnetic interactions. In addition, everything in the universe rotates because the photon rotates or spins, so any agglomeration of photons will rotate and since gravitational energy is also photons, any increase in gravitational energy, which then increases the magnitude of a gravitational force, since the gravitational interaction is dependent on energy, will lead to an increase in rotation, and a decrease in energy will lead to a decrease in the rotational speed of an object.

In conclusion, Planet X observations lead to a model for matter, energy and all interactions in the universe based on the photon or light, which already contains everything necessary within it, to give rise to all that we observe in the universe. In this way, light or photons are the building blocks out of which all matter and energy is made of in the universe. The model also explains the origin of the human aura since all isolated concentrations of photons are surrounded in a negative electric field and thus all isolated objects, at all scales, including the atomic scale, have an aura.

References:

[1]   Albers, C. and C'one, S. (2018). Book 3: Planet X revealed Gravity and Light.

[2]   Albers, C. (2019). Article 818: Planet X observations: the electrical universe is made of light.

Chapter 25

1182. Electrical density and electrogravitic drives

As I have shown in Article 1180: The photon model: all isolated objects have an aura [1], the gravitational interaction causes all isolated agglomerations of matter, to have an aura or an electric field around them. This is due to the gravitational interaction causing negatively charged plasma to be pushed outwards toward the outside surface of an object, which then makes the inner part to be positively charge, thus turning all isolated objects into capacitors.

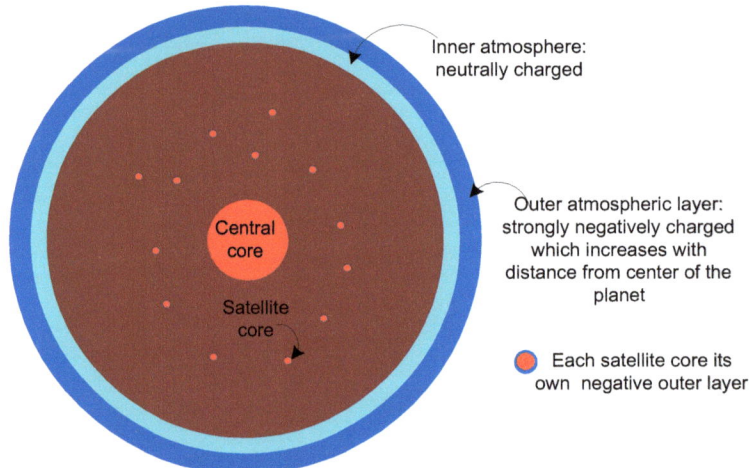

Figure 25.1. A planet will have an outer negative layer but so will each core. Thus, all cores will be surrounded in an electric field as will all agglomerations of photons, not moving at light speed, i.e. all atoms and all molecules will be surrounded in an outer electric field, produced by electrons. The planet's outer negative layer is the central core's negative outer layer.

But, since gravitational potential increases as we approach the center of the planet, matter closer to the center of the planet will tend to be more positively charged on the inside, than matter closer to the surface, or even outside the surface, whilst density also increases as we approach the center of a planet. In addition, density of matter in the atmosphere, the outer envelope of the planet, has a much lower density than matter inside the body of the planet. Furthermore, matter at the surface seems to be neutrally charged ,whilst matter in the atmosphere, at least at higher altitudes is negatively charged, which suggests that there is not only a relationship between gravitational potential and density there is also a relationship between electrical potential and density.

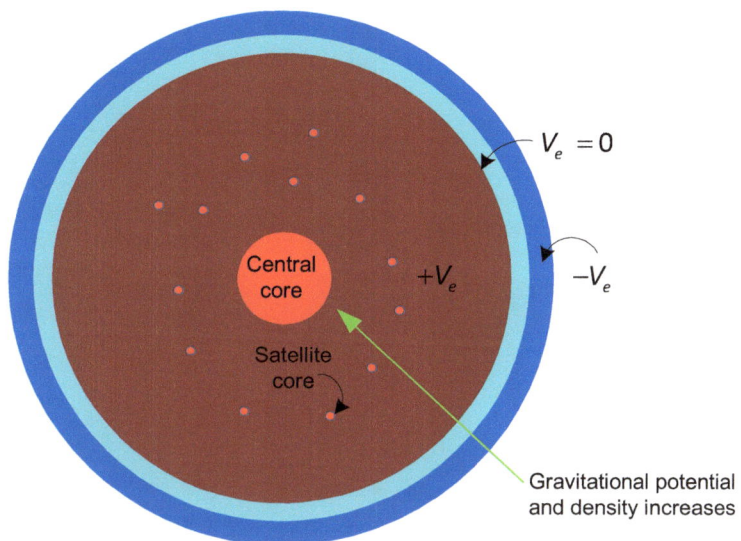

Figure 25.2. Gravitational potential and density are clearly related and increase in the same direction as we move toward the center of the planet. But there is also a relationship between electrical potential and density, although it is a bit more complicated.

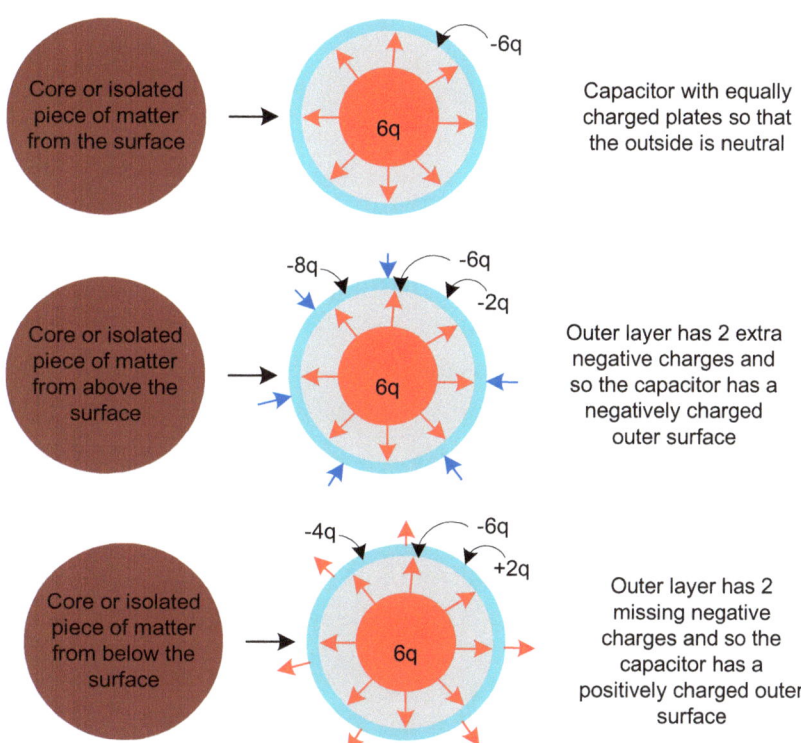

Figure 25.3. Cores from different parts of the planet: the surface, the inner part and from the atmosphere are turned into capacitors by the gravitational interaction, which have a different outer charge or field. A positive charge is associated to a positive potential and a negative charge is associated to a negative potential. Surface matter has a neutral outer surface, even though there is more negative charge on its outer layer than on the inside. An object from the atmosphere will have a negatively

charged outer surface and an object from deep within the planet will have a positively charged outer surface.

Since matter from deep within the planet has high density, we see that matter, which is positively charged, on its outer surface, will have a higher density, and matter, which has a negatively charged outer surface, will have a lower density.

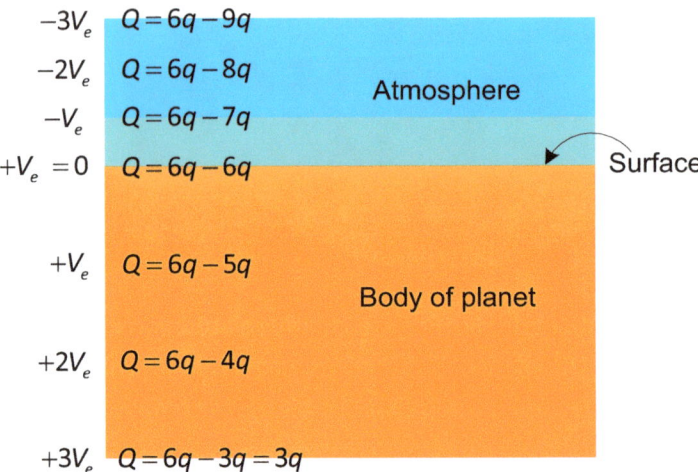

Figure 25.4. Potential and illustrative charges on the different parts of the matter which turn it into a capacitor: Matter in the body of the planet has an overall positive charge and matter in the atmosphere has an overall negative charge.

Matter also progressively decreases in density as we move up through the atmosphere. But Planet X debris that is of high density as a normal planet's matter goes, is observed to remain suspended in the atmosphere for prolonged periods, slowly sinking into the atmosphere, as it absorbs energy, which suggests that an object's position in the atmosphere has more to do with gravitational potential than density. In other words objects have an effective density which can be much less than their matter density which is dependent on gravitational energy or potential. But since gravitational potential is directly associated to electrical potential, electric potential is associated to the effective density of an object. And, since it is possible to negatively charge the outer surface of an object with an electrical power generator, it is possible to artificially decrease its gravitational potential or effective density so that it hovers in the atmosphere at a certain altitude. This is therefore the principle behind electrogravitic drives.

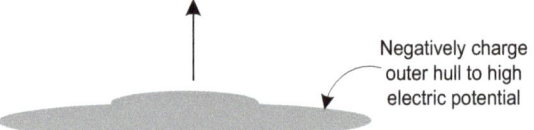

Figure 25.5. By taking a disk and charging the outer hull, metallic hull, whilst isolating it from matter in the interior to a high negative charge or potential, its effective density or gravitational potential will decrease, so that it rises within the earth's atmosphere.

Figure 25.5. Large Planet X debris pieces hovering in the atmosphere due to being gravitational energy depleted (see Article 1134: Intelligent life all over the Universe) [2]. The energy they absorb slowly allows them to turn into capacitors with an outer charge appropriate for matter on the surface of the earth, i.e. neutral outer surface. This also suggests that energy depleted isolated matter will not be positively charged, as I previously thought, but it simply lacks the energy to turn into a capacitor, as normal matter would do, thus completely energy depleted matter will be neutral and have no outer negative layer at all, it will be of uniform charge and potential throughout.

In conclusion, it has now become possible with the help of Planet X observations to understand how electrogravitic drives work. By charging the outer surface to a high negative potential its effective density or gravitational energy decreases, so that it rises within the earth's atmosphere. Thus electrogrivitic drives are based on electrical density.

References:

[1]   Albers, C. (2019). Article 1180: The photon model: all isolated objects have an aura.

[2]   Albers, C. (2019). Article 1134: Intelligent life all over the Universe (Book 13).

Chapter 26

1184. Static electricity on human body generated by the human creative spirit

It is common for static electricity to build up on the human body in winter, and especially when the humidity is low. This then leads to a spark occurring whenever a metallic object is touched as the buildup of charge transfers to the metal. We have been told that this buildup of charge is due to our shoes scraping across carpets or other types of insulating materials. But the human body will produce its own static electricity without any charge having to transfer to it from other materials. As I have shown in Article 1180: The photon model: all isolated objects have an aura [1], all isolated objects will automatically produce a negative outer layer, which results in an electric field on the outside of the object, due to that matter's gravitational energy or potential, which is in the form of photons or light. The human body is no different and will thus produce an electric field outside the body, i.e. an aura, due to the additional charge on the outside surface of the skin.

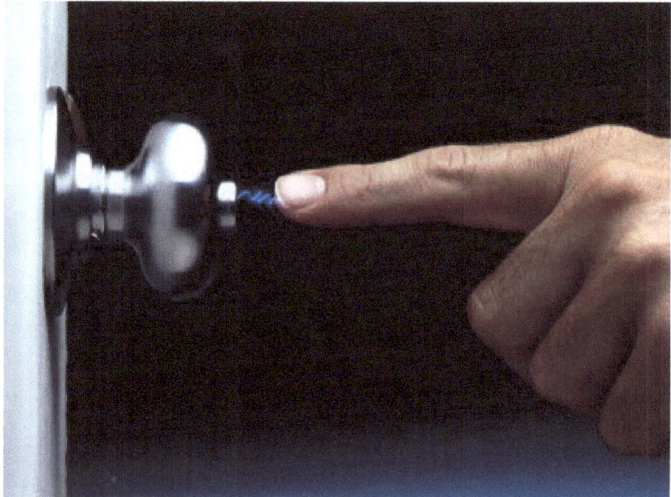

Figure 26.1. A buildup of charge on the outside of the human body causes a discharge whenever the body closely approaches a metallic object. This occurs before the object is even touched indicating a build up to a very high electric potential on the outer skin surface, as the charge transfers across an air gap, which shows that the human body has a powerful ability to generate gravitational energy or light, on the inside.

The reason why a spark tends to happen, more frequently in winter and under low humidity conditions, when a metallic object is approached, is that water vapor in the atmosphere transports the extra charge away from the human body, but when there is very little water vapor in the atmosphere less charge is transported away from a person's skin and therefore the charge accumulates. However, what happens in winter and low humidity conditions indicates that a high amount of charge is constantly building up on the outside of the human body, due to a high degree of gravitational energy being generated on the inside. It is simply impossible for this huge potential to be due to rubbing one's shoes on the carpet. If this was possible, static electricity generators would be in use all over the world that used carpet and plastic soles to generate charge.

Thus, the human body is a source of the static electricity and a very strong source at that. So what makes a human being such a high generator of gravitational energy or light? It must be that a human being has a spirit which constantly generates energy or light. A human being has a spiritual core, with creative ability, and is thus a dynamo of spiritual energy. This spiritual energy is what God used to create the universe and is what powers the universe and this energy is light or photons (see Article 1144: The Universe is alive: God is reproducing Himself, Article 833: Planet X shows that the Universe works according to Biblical principles and Article 818: Planet X observations: the electrical universe is made of light) [2, 3, 4]. This means that light is spiritual energy and that we are living in a light and thus a spiritual universe and that we ourselves generate that light inside of ourselves and it manifests as an electric field or aura around our bodies.

Physical cores absorb energy from other cores but creative beings generate that energy from their spirits. Human beings are created in God's image and so just like God can create using the creative energy, which His Spirit creates, so can human beings. Thus, the constant buildup of charge on the human body is a manifestation of energy being generated by the human spirit. However, God's ability to create energy is unlimited, whilst ours is limited, until we get connected with Him by getting born again. Lucifer got us disconnected from our Father in the Garden of Eden, but we can now get connected again by asking Him, Jesus, to take our lives.

Figure 26.2. Planet X System Stellar Cores (SCs) or the inside parts of the Planet X planets inside the earth's atmosphere and surrounded in cloud envelopes. All cores are clothed with envelopes that emit light, the human aura is a human being's outer envelope.

In conclusion, the human spirit generates spiritual energy or light which manifests as a constant buildup of charge on the human body's outer layer of skin which then gives rise to the human aura, which is a human being's outer envelope or clothing. This energy generated by the human spirit is made of photons or light and is also spiritual energy and is also the same energy out of which the whole universe is made of. We are therefore not living in a physical universe contained inside a spiritual universe; we are living inside a spiritual or light universe.

References:

[1]   Albers, C. (2019). Article 1180: The photon model: all isolated objects have an aura.
[2]   Albers, C. (2019). Article 1144: The Universe is alive: God is reproducing Himself.
[3]   Albers, C. (2019). Article 833: Planet X shows that the Universe works according to Biblical principles.
[4]   Albers, C. (2019). Article 818: Planet X observations: the electrical universe is made of light.

Chapter 27

620. Huge gas giant planets in the inner Solar System

As I have shown in previous articles Planet X System Stellar Cores (SCs) which come to the Sun turn into gas giant planets, which orbit the Sun (see Article 523: Planet X and the Solar System: Jupiter and all gas giants are recent acquisitions) [1], whilst SCs, which come to the earth, turn into earth moons (see Article 526: Planet X and the Moon: the Moon has not always been in sky) [2]. Jupiter is about one tenth the size of the Sun and it ended up orbing at a distance of 5.2 au away from the Sun. Much larger SCs have come into the Solar System and have been observed in the Sun's corona, which would have gone through the same reenergizing process that the SC that became Jupiter did, i.e. by absorbing electrons and matter from the Sun.

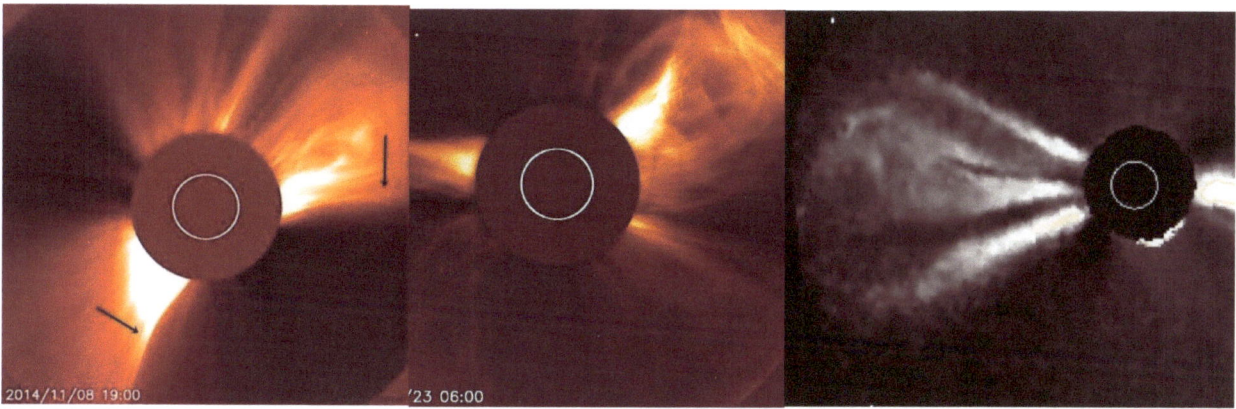

Figure 27.1. Three Stellar Cores: A: 7 times the size of the Sun, B: about the same time the size of the Sun and C: 4 times the size of the Sun.

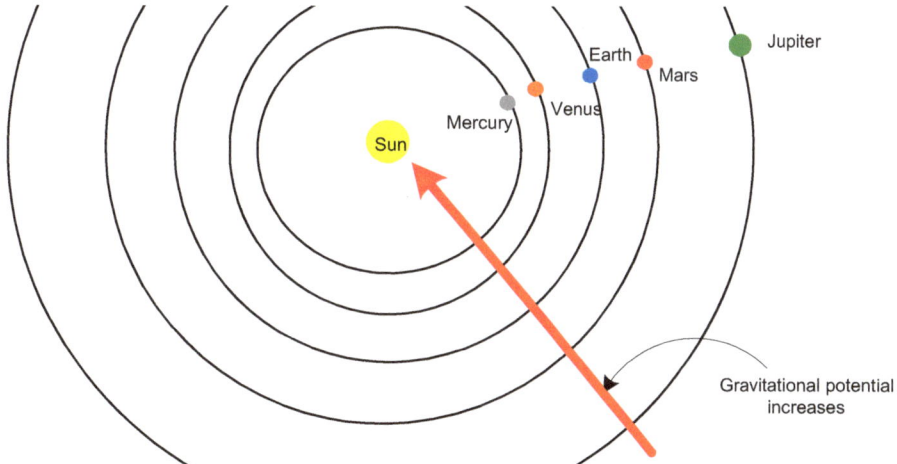

Figure 27.2. As gravitational energy increases, in the direction of the Sun, the highest gravitational energy objects will end up orbiting closest to the Sun and the largest SCs will end up with the most gravitational energy and thus orbiting closer to the Sun than smaller SCs.

This means that there must be an inverse relationship between size and orbital distance but we do not know to what power. In order to determine what the real relationship is between the size of the gas giants and their orbital radius, we will use the orbital parameters for the 4 known gas giants:

Gas Giant	R (R_{Sun})	r (au)
Jupiter	0.1	5.2
Saturn	0.0837	9
Uranus	0.0365	19.2
Neptune	0.0354	30.1

where r is orbital distance and R is radius of the object. So the relationship equation is given by:

$$R = r^x \Rightarrow \ln R = x \ln r$$

We need to determine x from the slope of the graph we get by plotting ln$R$ versus ln$r$, which is shown below:

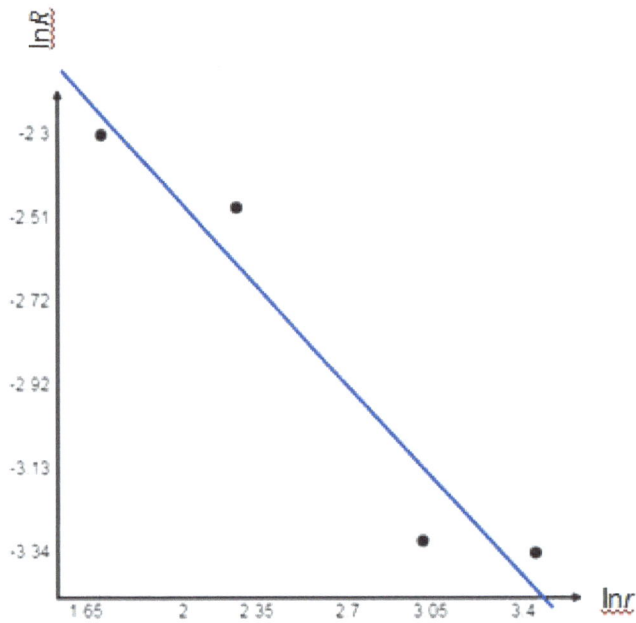

Figure 27.2. The slope to the graph is -0.675, which is very close to -2/3, so we will take that to be the relationship between the radius and the orbital radius of the gas giants.

The fact that the number is so close to 2/3 is strong evidence that the observation based theory, which led to the understanding that gas giants are re-energized Stellar Cores (SCs) and that smaller SCs would orbit further from the Sun, is correct.

$$R \propto r^{-2/3} = \frac{1}{r^{2/3}} \Rightarrow r \propto \frac{1}{R^{3/2}}$$

Now, the SC which is about the same size as the Sun would have a radius $R_{SCB} = R_{Sun}$ and for Jupiter we have: $R_J = 0.1 R_{Sun}$ and $r_J = 5.2$ au. Thus,

$$\frac{r_{SCB}}{r_J} = \left(\frac{R_J}{R_{SCB}}\right)^{3/2} \Rightarrow r_{SCB} = \left(\frac{0.1 R_{Sun}}{R_{Sun}}\right)^{1.5} 5.2 \text{ au} = 0.16 \text{ au}$$

So, this SC would become a gas giant planet of about the same size as the Sun orbiting about between Mercury and the Sun, as Mercury has an orbital radius of 0.39 au. Then, the Stellar Core C, which was 4 times larger than the Sun would end up at:

$$\frac{r_{SCC}}{r_J} = \left(\frac{R_J}{R_{SCC}}\right)^{3/2} \Rightarrow r_{SCC} = \left(\frac{0.1 R_{Sun}}{4 R_{Sun}}\right)^{1.5} 5.2 \text{ au} = 0.02 \text{ au}$$

And the Stellar Core A, which was 7 times larger than the Sun, would end up at:

$$\frac{r_{SCA}}{r_J} = \left(\frac{R_J}{R_{SCA}}\right)^{3/2} \Rightarrow r_{SCB} = \left(\frac{0.1 R_{Sun}}{7 R_{Sun}}\right)^{1.5} 5.2 \text{ au} = 0.0089 \text{ au} \simeq 0.01 \text{ au}$$

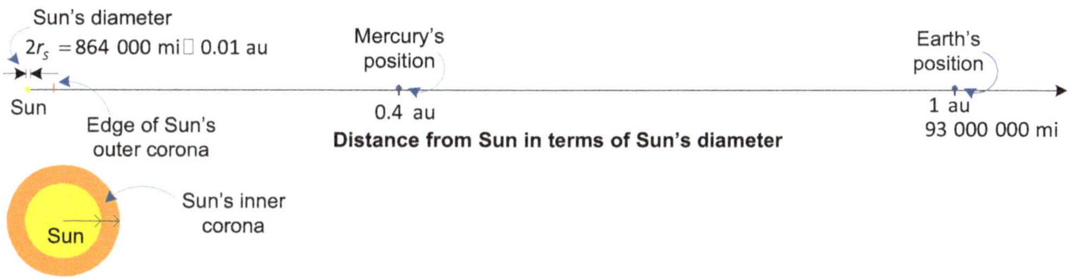

Figure 27.3. The Sun's outer corona goes out to a distance of 0.06 au, so the two larger Stellar Cores would end up well within the Sun's outer corona. In fact their orbital radius is so close to the Sun that they would actually have to be inside the Sun because of their size which is not possible.

Since these objects are larger than the Sun they would have to actually go inside the Sun in order to orbit at these close distances, which is not possible, since they will be repelled by the Sun's core, which will thus push them outwards from the Sun, until their surfaces are some distance away from each other. So the closest distance that SCA can orbit the Sun is at 0.04 au.

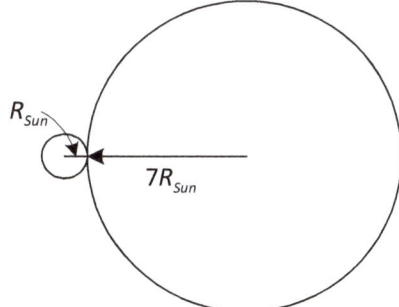

Figure 27.4. SCA can only be as close to the Sun as $8 R_{Sun}$ and since $R_{Sun} \simeq 0.005$ au, the object cannot be any closer than 0.04 au to the Sun.

Now, we can calculate the period of the orbit using Kepler's third law:

$$T \propto r^{3/2}$$

Assuming that r_{SCA} = 0.04 au, since we know that r_{earth} = 1 au and T_{earth} = 365 d. Thus

$$\frac{T_{SCA}}{T_{Earth}} = \left(\frac{r_{SCA}}{r_{Earth}}\right)^{3/2} \Rightarrow T_{SCA} = \left(\frac{0.04 \text{ au}}{1 \text{ au}}\right)^{3/2} (365 \text{ d}) = 2.9 \text{ d}$$

So, this gas giant 7 times larger than the Sun would orbit the Sun once every 3 days. Stellar Core C, which is 4 times the size of the Sun would not be able to get any closer than 0.025 au to the Sun and at that distance would orbit the sun once every 1.4 days, then SCB, which is about the same size as the Sun, will become a gas giant orbiting the Sun with an orbital period of 23 days.

Hence SCA would have an orbital velocity of

$$v_{SCA} = \frac{2\pi r_{SCA}}{T_{SCA}} = 3.611 \times 10^{-3} \text{ au/h}$$

Then, we can calculate how long this object would eclipse the Sun as it passes in front of it once every 3 days.

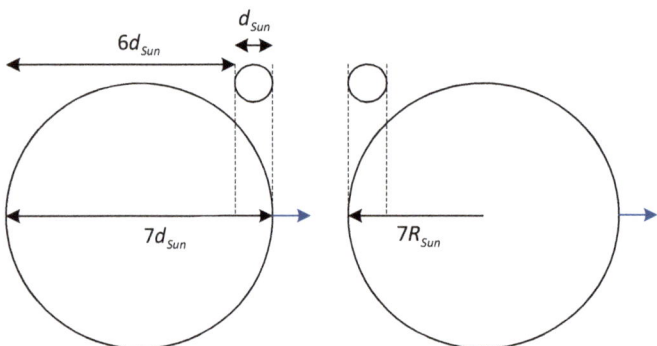

Figure 27.5. The Sun would not be visible as the object moves a distance of $6d_{Sun}$ = 0.06 au.

At its orbital velocity, SCA will take 16.6 hours to move a distance of 0.06 au, so that would mean that the Sun would not be seen from earth for nearly 17 hours every 3 days. This is if it was shining at all, as it does not seem to be, as it is never seen from earth (see Article 500: The sun is no longer shining review) [3] and was already not being seen at all by 1998 (see Article 536: Was the Sun already not shining by 1998?) [4]. The question then is, with such large gas giants orbiting the Sun how are the SDO images, Stereo images, SOHO images even possible? Most likely the images we are been shown are very old images and if there were large gas giants already orbiting the Sun at the time that these images were being captured, most likely something was done to hide the fact that the Sun was being eclipsed. Either way some heavy doctoring has most likely being done to them.

In conclusion, there must now be quite a few very large gas giant planets orbiting the Sun which would be eclipsing the Sun on a regular basis even if the Sun was still shining if the solar images showing these objects close to the Sun are true. There is however a distinct possibility that these images are either simulated or manipulated so that Planet X star cores much larger than they really are appear in them.

References:

[1] Albers, C. (2018). Article 523: Planet X and the Solar System: Jupiter and all gas giants are recent acquisitions.
[2] Albers, C. (2018). Article 526: Planet X and the Moon: the Moon has not always been in sky.
[3] Albers, C. (2018). Article 500: The sun is no longer shining review.
[4] Albers, C. (2018). Article 536: Was the Sun already not shining by 1998?

Chapter 28

621. Gas Giant vindication and why Dr Eugene Shoemaker was really killed

In Article 620: Huge gas giant planets in the inner Solar System [1] I showed how the known Solar System Gas Giant planets, very neatly, follow a relationship between their orbital distance from the Sun and their radius, i.e.

$$r \propto \frac{1}{R^{3/2}} \tag{1}$$

where *r* is orbital distance and *R* is the radius of the Gas Giant. It is obvious that the other planets, in the Solar System, do not follow this relationship as both Mercury and Venus are smaller than earth, and they would have to be much larger than earth, if they did. This is because the outside radius, of the rocky planets, is different from the size of their core. This only works in the case of the Gas Giants because they have a very thin layer of gas, on top of their cores, so their overall radius is approximately the same as their core's radius.

Figure 28.1. A Planet X System Stellar Core in the process of becoming a gas giant as evidenced by its stripes. This object cannot have yet finished the re-energizing process because it is very close to the Sun and is also much smaller than the Sun. Only gas giants much larger than the Sun end up orbiting the Sun, this close to it, as I showed in Article 620, i.e. a Stellar Core 7 times larger than the Sun will end up orbiting at 0.04 au from the center of the Sun [1]. The SC has a trail of gaseous plasma behind it, indicating that it is absorbing gaseous plasma from the Sun's corona.

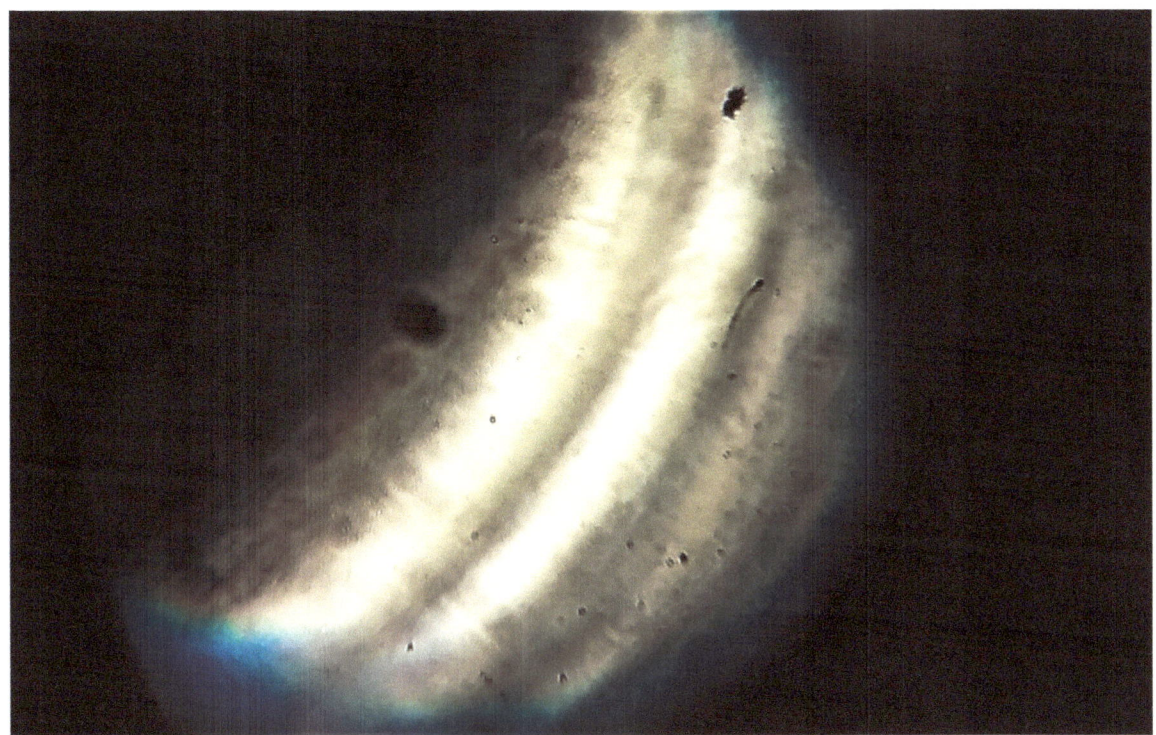

Figure 28.2. Another Planet X System Stellar Core photographed by an astronomer and sent to Scott C'one, also, most likely, in the process of becoming a gas giant, as its gaseous material seems to be in widely varying colors, i.e. blue on the edge and beige or brown on the side facing the camera. This suggests that its covering is changing from most likely the cloud envelope it arrived, at the Sun, with, to a new covering, which it is absorbing from the Sun, which will give it a striped appearance like that of Jupiter.

Thus, since their overall radius is about the same as the core's radius, these objects have a very thin layer of gaseous material, which they absorb from the Sun's corona, and therefore Jupiter, also has a very thin layer of gas and is mostly a solid object. This is why the comet Shoemaker-Levy 9 impact on Jupiter left such lasting impressions on the planet. The comet pieces did not impact a gaseous object, they impacted a solid object, with a very thin layer of gas, on top of the solid surface. This is therefore the reason why Dr. Eugene Shoemaker, the planetary scientist, who was also the co-discoverer of the comet, was killed. He knew from the impacts that Jupiter was solid, almost all the way to the top of the gas clouds.

Figure 28.3. The different pieces that comet Shoemaker-Levy 9 broke into.

Figure 28.4. Image of Jupiter, in ultraviolet light, after the impacts, showing the large blemishes that were left on Jupiter, which would be impossible, unless there was a solid surface right underneath the top of the gaseous layer.

Now, my understanding that the Gas Giants are re-energized Planet X System Stellar Cores came through observing these objects and then determining how gravity works and how the laws governing the universe work. So, it was the developed theory that led to my believing that gas giant planets' orbital distance had to be related to their radii. Thus, to then find that the four known Gas Giants' orbital parameters fall within such a neat mathematical law, as indicated by equation (1), which is similar to Kepler's third law, in terms of simplicity:

$$T \propto r^{3/2} \text{ (Kepler's third law)} \quad \text{and} \quad r \propto \frac{1}{R^{3/2}} \text{ (orbital radius to core radius relationship)}$$

where T is orbital period, r is orbital radius and R is the radius of a celestial object's core, is a real verification of my theories as well as vindication. The thing is; this data has been around for thousands of years and astronomers, astrophysicists have had enough understanding of the universe to make use of it for centuries. This relationship clearly tells us that Gas Giants have to be mainly made of a core that is the same density, for all of them, and that they are different from all the other planets in the Solar System. Why hasn't anyone figured this out until now? Why are there no published papers on this? Because the physics and astronomy on this planet have been subverted and controlled by the father of lies and those people who discovered the truth and cannot be silenced, or stopped from spreading the truth to the rest of the community, by being ejected from academia, were killed like Dr. Eugene Shoemaker. This also shows how the understanding of what these objects really are, what the Planet X System is, is something that Lucifer wants to hide from the earth's human population at all costs.

However, Lucifer is not really in control, God is in control, and Lucifer can only do what God allows, which why the Creator God has seen fit to reveal these truths, at the end of this age, through me and many other of His faithful children.

In conclusion, the relationship between orbital radius and radius of the Gas Giant planets is a form of verification for the theory I have come up with, through observing Planet X System Stellar Cores, and vindication in the face of opposition from the father of lies; it also explains why scientists like Dr. Eugene Shoemaker were killed and, in addition, shows that the truth about Planet X is something that Lucifer would like to remain hidden at all costs.

References:

[1]   Albers, C. (2019). Article 620: Huge gas giant planets in the inner Solar System.

Chapter 29

103. The 800 million solar masses black hole at the edge of the universe

On December 7th 2017, a story appeared in the major scientific websites geared to the public, such as space.com and phys.org about a massive black hole at the edge of the universe, with mass which is 800 million times more than our Sun [1]. The article was based on an astrophysical letter published in the journal Nature, and available under their accelerated article preview program regarding the quasar ULASJ134208.10 + 092838.61. The article refers to a quasar of redshift 7.5.

Figure 29.1. Artist's impression of a Black Hole. The twisting in the beam of seeming to be emitted by the object looks like a Birkeland current and which correctly refers to the electric nature of the object.

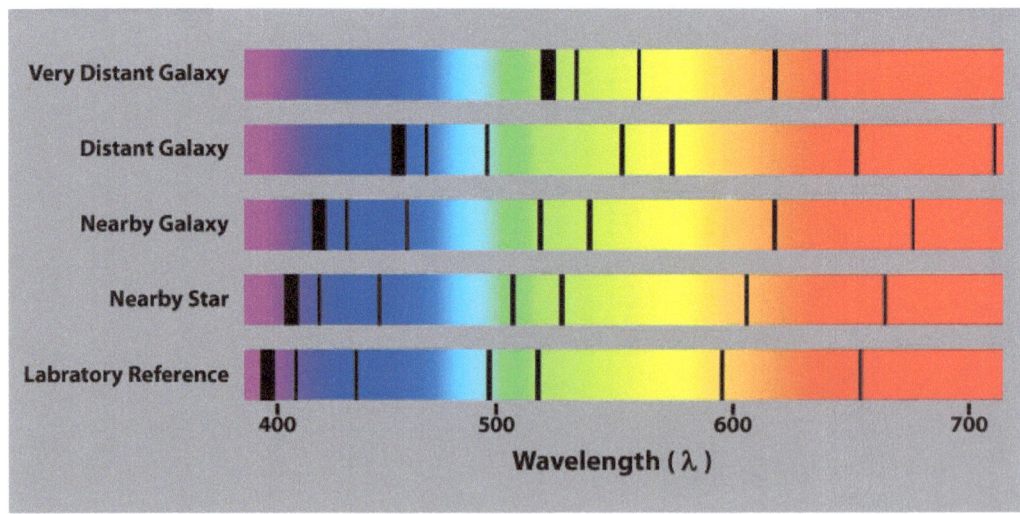

Figure 29.2. Illustration of redshift: the lines in an astronomical object's spectrum are shifted toward the red end of the spectrum.

Now, redshift refers to the shifting of an astronomical body's light spectrum towards the red side of the spectrum. Redshift is based on the Doppler Effect which explains the difference in the pitch of the sound emitted by an object, as it moves with respect, to an observer listening to this sound. The sound of an ambulance's siren is different when the object approaches, to what it is when the object speeds away. When the object speeds away from the observer, the frequency of the sound decreases. This also applies to light, so that light emitted by an object which is moving away from the earth, would have a higher wavelength, and thus its spectrum would seem to have shifted toward the red end of the spectrum, as illustrated above. This means that the light emitted by the object would have a lower frequency than the light emitted by the object, if it were not moving with respect to the observer, i.e. earth.

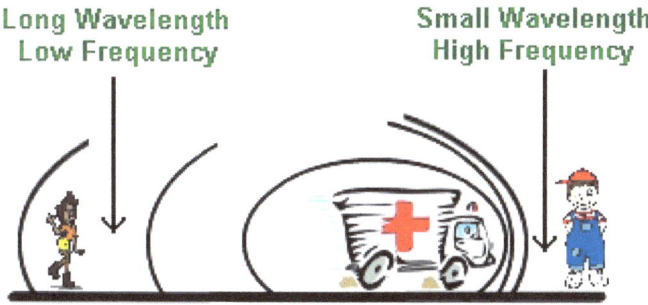

Figure 29.3. The wavelength of a sound wave increases (frequency drops) as an object moves away and the wavelength decreases (frequency increases) when it approaches.

The idea that redshift is due to an object moving away from earth provides a way of calculating the object's velocity. Objects with the highest redshift would thus be moving faster. The objects that are moving the fastest are also thought of as being further away from earth. This is based on the Hubble Law which states that the speed with which an astronomical object is moving away from earth is directly proportional to its distance from earth. This is the law on which the Big bang model is based.

Figure 29.4. The universe contains at least 100 billion galaxies.

The Big Bang model is based on the fact that because it takes a long time for light to get to earth from objects that are very far away, these objects are seen by us as they looked a very long time ago. So an object which is 1 light year (6 trillion miles) away would be seen, by us, as it looked 1 year ago, and an object at a distance of 1000 light years (6 quadrillion miles) away would be seen as it looked like 1000 years ago. Thus, the Hubble Law basically leads to the idea that the further back in time we look, the faster the objects we observe seem to be moving and that therefore the universe is expanding, and that there was a time when it had zero volume, at which time matter was moving the fastest or exploding. This has several problems: first of all, something with zero volume is undefined or does not exist. So if the universe ever had zero volume then it did not exit. Secondly, creating space and matter out of something that does not exist contradicts the laws of thermodynamics, on which the whole of physics is based, and which states that matter or energy cannot be created or destroyed.

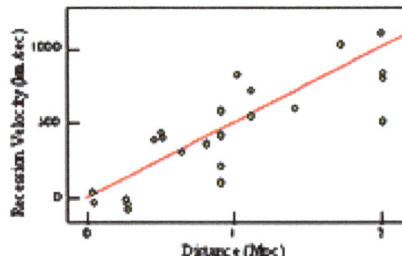

Figure 29.5. Hubble's Data has been extrapolated in a completely unreasonable manner, in order to yield the Big bang Model, which makes no physical sense, has been falsified and should have been abandoned a long time ago.

But going back to redshift, based on the idea that redshift is all due to the velocity with which an object is moving away from earth. The equation which relates red shift to recessional velocity is

$$v = c\frac{(z+1)^2 - 1}{(z+1)^2 + 1}$$

Where v is the recessional speed, c is the speed of light and z is the redshift. Thus, an object with a redshift greater than 7 is expected to be moving away extremely fast, and very close to the speed of light (0.97c or 290 000 km/s) and then base on the Hubble law, the object would have to be extremely far away, and also in the extreme past or very close to the beginning age of the universe.

The problem is that most of the value associated with the redshift of an object cannot be due to the speed with which it is moving away from earth. In his book 'Seeing Red: Redshifts, Cosmology and Academic Science', the astronomer Halton Arp details many instances when high redshift objects are at exactly the same distances from earth as low redshift objects. The book is full of detailed explanation around astronomical images where high redshift objects like quasars are connected by dust, to low red shift galaxies. How can an object which is supposed to be at the edge of the known universe and therefore way in the past be connected to a galaxy that is supposed to be relatively close to our own? It is impossible. Halton Arp's work has completely falsified the Hubble Law and the Big Bang Model and both of these should have been abandoned by astrophysicists, a long time ago, but we continue to see this type of work continue to be published. In fact, it was so 'acceptable' that it was available under the accelerated article preview program.

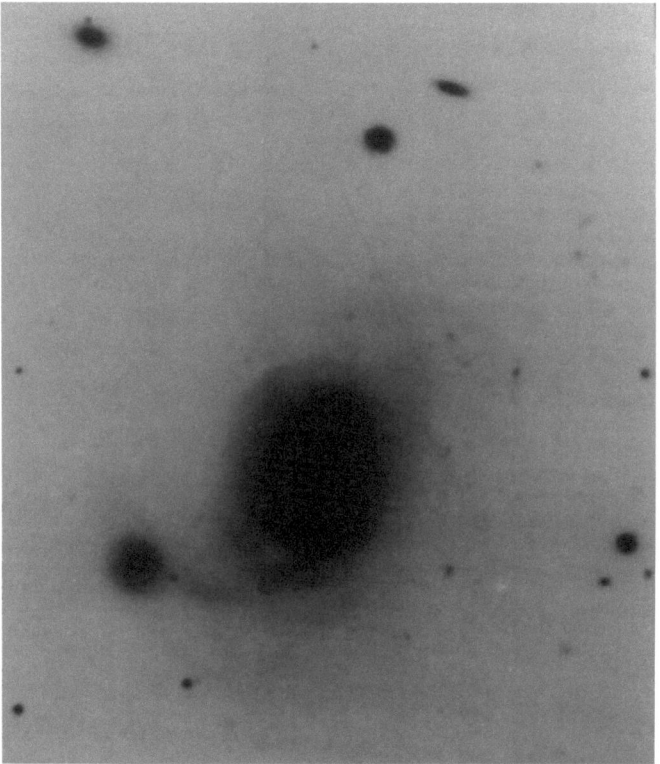

Figure 29.6. Image on page 112 of Seeing Red, by Halton Arp: The Seyfert galaxy NGC7603 is connected by dust to a smaller companion. NGC7603 has a redshift equivalent to a speed of 8 700 km/s and its companion has a redshift equivalent to a speed of 17 000 km/s, or nearly twice the velocity, which according to the Hubble Law should place it twice as far away from earth. However, since the objects are connected they have to be at the same distance from earth, which falsifies the Hubble Law and with it the Big Bang Model.

What Halton Arp found is that the greatest percentage of the measured redshift of an object is intrinsic and only a small percentage may be due to speed, with respect to the observer. He also found that objects like quasars are ejected from active galactic nuclei, or from the nucleus of galaxies, that are particularly bright and thus emitting a lot of radiation, and that these objects have very high redshifts. Also the younger the quasars are the higher their redshift. That is, the closer they are to the parent galaxy, the higher their redshift seems to be. This means that redshift has to do with the age of the object. Thus, the redshift of quasars decreases as they age and this decrease happens in steps or is quantized. Halton Arp also found that as they age quasars become larger and unfurl arms and thus become young galaxies. This means that galaxies give birth to galaxies and that matter is created within the extremely bright and energetic environment of the nucleus of galaxies and not in a huge Big Bang from an object of zero volume that could not have existed.

In fact, Halton Arp writes, in Seeing Red, that because objects have an intrinsic redshift associated to their age, it means that mass increases as objects age. Thus, material is ejected from an active galactic nucleus, which is made up of particles that have near zero mass and that the mass of the particles increases in steps as the material ages. This makes sense because an object with a very high redshift is emitting photons, with a very low frequency, and therefore energy, which should be associated to particles of much lower mass, than those in our galaxy.

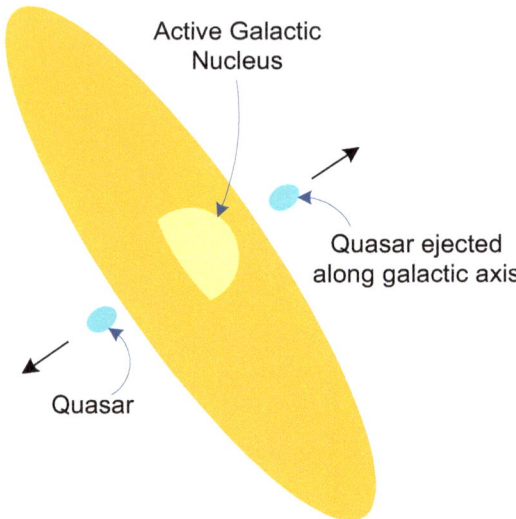

Figure 29.7. Active Galactic Nuclei are intensely bright galactic centers, with extremely high electric fields. When the electric field and brightness reaches a critical level, it causes instability and the ejection of matter in the form of a quasar. In the course of time the quasars unfurl arms, as they start ejecting their own material, and thus become galaxies themselves. Thus galaxies give birth to galaxies and matter in continuously being created in the universe.

Now the mass of the quasar in the letter in Nature was determined from a study of its luminosity, which again ignores the fact that mass itself is dependent on redshift, and the fact that we are living in an electrical universe not a gravitational one. Observation of the Stellar Cores in the Sun's Corona has revealed that the universe is electrical [2, 3] in nature. In fact, gravitational collapse fails to explain both the formation of star systems [4] and the workings of pulsars [5]. Thus, gravitational collapse constructs such as neutron stars and black holes cannot exist. Instead of a huge mass, a high luminosity object is one with an extremely high electric field. Very bright galaxies are called active because they are known to eject material and to emit a lot of radiation, and this happens as a result of the extreme electric field generated by the nucleus of such galaxies, which becomes unstable to the point that matter is ejected. It is the same process that leads to cosmic ray bursts and solar flares [6].

In conclusion, based on the observation of Stellar Cores, in the Sun's corona, and the work done by the astronomer Halton Arp, the quasar, ULASJ134208.10 + 092838.61, which is a subject of this astrophysical article cannot be a black hole as black holes cannot exist in an electrical universe. This quasar is also neither very old nor at the edge of the universe, as its extreme redshift is due to its young age rather than due to recessional velocity.

References:

[1]   Banados, E. et al. (2017). An 800 million solar mass black hole in a significantly neutral universe at a redshift of 7.5. *Nature* http://dx.doi.org/10.1038/nature25180.

[2]   Albers, C. (2017). Article 67: Stellar Core Gravitation and Eris.

[3]   Albers, C. (2017). Article 73: Newly Discovered Exoplanet Planet X Stellar Cores and the Electrical Universe.

[4]   Albers, C. (2017). Article 95: Planet X reveals the correct Star System formation model.

[5]   Albers, C. (2017). Article 83: The Neutron Star impossibility.

[6[Albers, C. (2017). Article 75: Planet X Stellar Core composition and Gamma Ray Bursts.

Chapter 30

126. White Holes instead of Black Holes at the Center of Galaxies

Black Holes cannot exist as gravitational collapse cannot be the driving mechanism behind stellar formation, and without gravitational collapse, Black Holes, which are singularities, or point sized objects, with infinite gravity, surrounded by an accretion disk, cannot form.

Figure 30.1. An artist's impression of a Black Hole: It is made up of singularity surrounded by an accretion disk of matter, which is falling into the Black Hole. According to Einstein's theory of General Relativity, it should take an infinite amount of time for matter to fall into a singularity, which makes the formation of a Black Hole impossible.

The evidence for the fact that gravitational collapse cannot occur comes from the observation of the system of Planet X Objects or Stellar Cores which have invaded the Solar System and that are often found in the Sun's Corona. These are old stars, which have aged beyond the point when they can emit visible light, but once in the Sun's corona they absorb energy from the Sun, they rejuvenate and are once again able to emit light. As discussed in previous articles, these objects do not interact gravitationally with the Sun; they do not collide with it and are at times repelled by it. In addition, the Sun goes dark at times, which makes it impossible for it to be powered by thermonuclear reactions, and thus cannot have formed as a result of gravitational collapse.

So if Black Holes cannot exist, why are accepted theory proponents suggesting that there has to be a supermassive black hole at the center of each galaxy? The reason for this lies in the way distances are determined, which is based on the measured redshift of cosmic objects. The redshift is the observed lines in the spectrum of these objects, which are shifted toward the red side of the spectrum, in relation to the spectrum measured in a laboratory on earth.

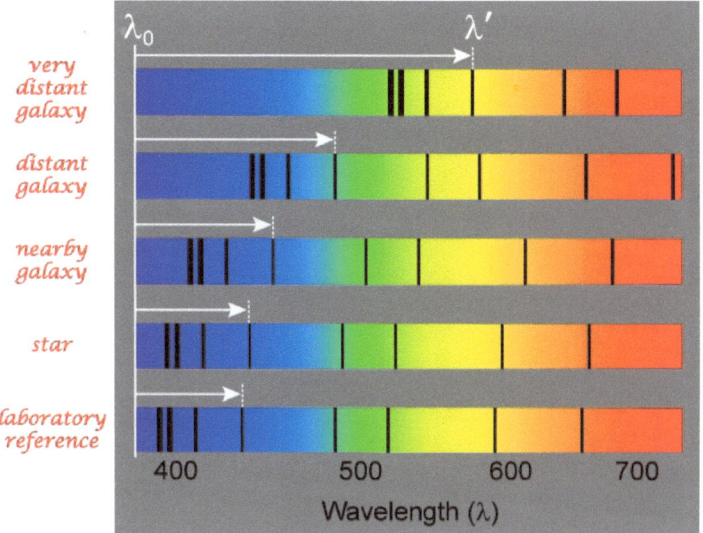

Figure 30.2. When redshift is taken to mean recessional velocity, the higher the measured redshift of an object the further it will seem to be.

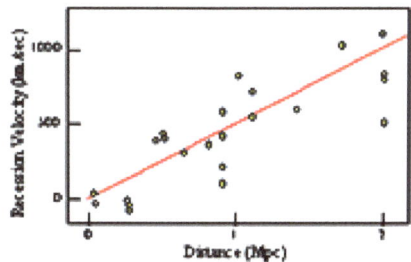

Figure 303. Hubble's Data has been extrapolated in a completely unreasonable manner, in order to yield the Big bang Model, which makes no physical sense, has been falsified, and should have been abandoned a long time ago.

The idea that redshift has to do with distance came from data obtained by Hubble in 1929, shown below. Data at the time was only available for a few nearby objects but it seemed to suggest that the redshift of objects, which were further away, was higher than for those closer to earth. However, Hubble himself was never totally convinced that redshift was due to recessional velocity, and until the end of his life, believed that it could be due to something else. The fact that it has to be due to something else is easily shown by the many photographs of high redshift objects connected by dust to low redshift objects, as shown in the image below from Halton Arp's book 'Seeing Red' [1]. When the recessional velocity to redshift connection is assumed, it is possible to simply write redshift in terms of that velocity. The redshift, of the objects shown in figure 4, is written in that form. For more details, see article 103: The 800 million solar masses black hole at the edge of the universe [2].

The Hubble Law is based on the interpretation of redshift as recessional velocity and states that the further back in time we look, the faster the recessional speed, and therefore that the universe is expanding, thus leading to the Big Bang Model [2].

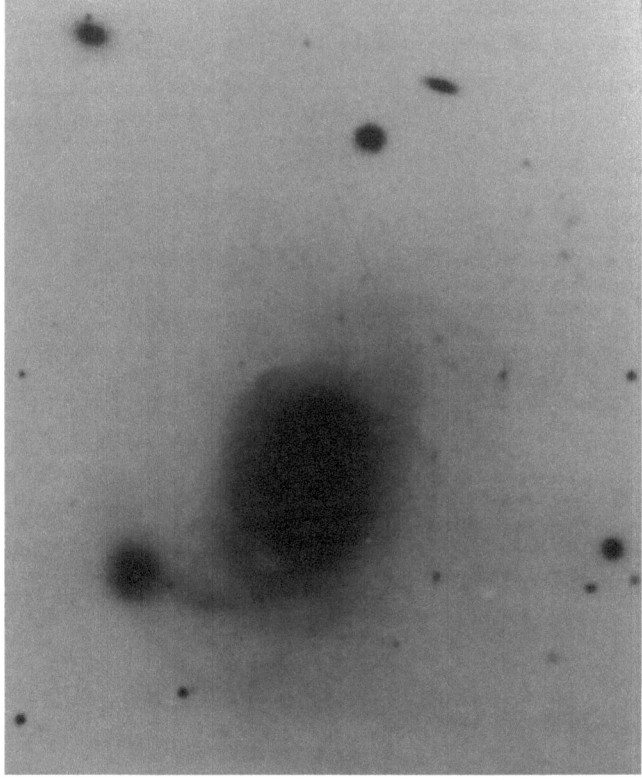

Figure 30.4. Image on page 112 in Seeing Red, by Halton Arp: The Seyfert galaxy NGC7603 is connected by dust to a smaller companion. NGC7603 has a redshift equivalent to a speed of 8 700 km/s and its companion has a redshift equivalent to a speed of 17 000 km/s, or nearly twice the velocity, which according to the Hubble Law should place it twice as far away from earth. However, since the objects are connected they have to be at the same distance from earth, which falsifies the Hubble Law and with it the Big Bang Model. Seyfert galaxies have extremely bright central regions or in other words Active Galactic Nuclei.

However, redshift was shown by Halton Arp, in his book 'Seeing Red' not to be related to recessional velocity but that it is an actual intrinsic property of matter, which has to do with the age of that matter. The younger the matter, the higher will be its intrinsic redshift. It has been known, since 1948, that galaxies eject material from their nuclei. This material emits radio waves. One example of this typical ejection can be seen in the galaxy Cygnus A, which emits high energy radio wave emitting material from its nucleus. The ejections are in opposite directions.

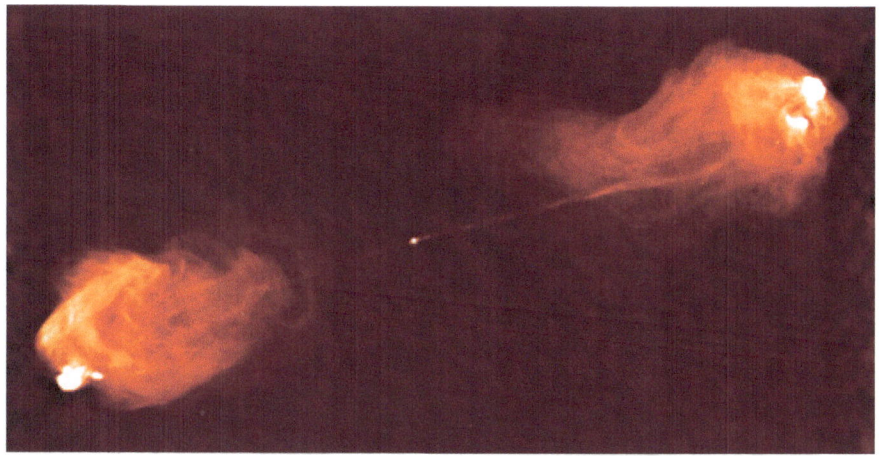

Figure 30.5. Cygnus A is a galaxy, which ejects high energy radio emitting material from its nucleus.

According to accepted theory there is a supermassive Black Hole at the center of Cygnus A, which is the source of this ejection. However, since nothing is supposed to be able to escape a Black Hole, how is it possible that this ejection of high energy material is occurring from this Black Hole? And why do we need to have a Black Hole at the center of the galaxy anyway? The reason for this is that because redshift is used to measure distances, galaxies will often appear to be further away than they really are. Now, a galaxy that is deemed to be further away, than it actually is, will seem to be brighter than it actually is, and thus its luminosity will be overestimated. Luminosity is the amount of energy, in a particular wavelength range, often in the optical or visible light range, emitted by an object per second.

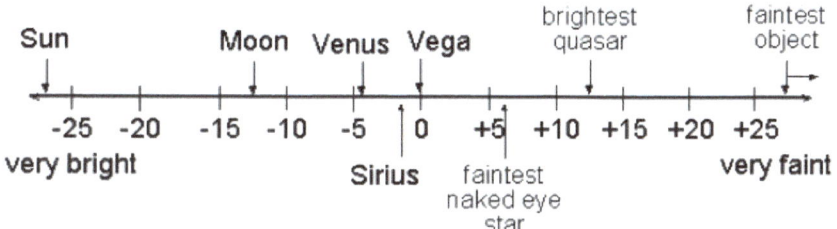

Apparent brightnesses of some objects in the magnitude system.

Figure 30.6. The apparent brightness, or magnitude, of several celestial objects: An object's mass can be estimated from its luminosity. An object's luminosity can be calculated from its apparent brightness if its distance from earth is known.

If two objects have the same luminosity, they are emitting the same amount of energy per second (power) and will have the same mass. But if two objects are equally bright, but one is twice as far as the other, than the one that is further away is more luminous than the one that is close, and must therefore be more massive than the object that is close. This is illustrated below.

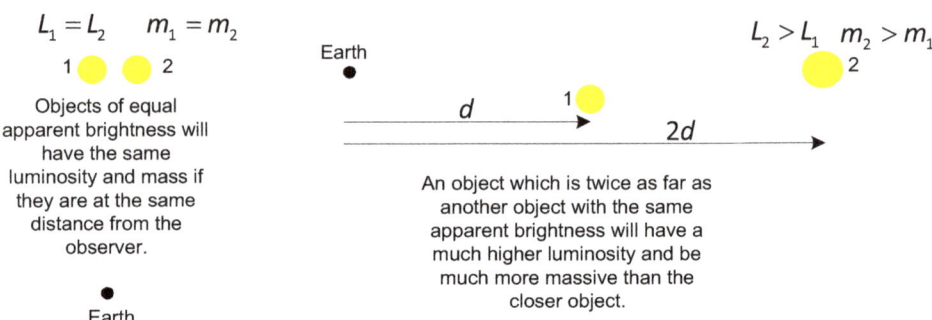

Figure 30.7. Illustration of the relationship between apparent brightness, luminosity and mass: Two objects at the same distance, and with the same apparent brightness, will have the same luminosity. Two objects with the same luminosity will then have the same mass. If an object seems to be just as bright as another but it is twice as far, than its calculated luminosity, and mass, will be higher than for the object that is close.

Thus, if a galaxy is relatively close to our own galaxy but its redshift is high, its distance from earth will be deemed to be greater than normal, and therefore the object's luminosity will be overestimated. In other words, a much higher than the true value, will be arrived at. Thus, its mass will also be much higher than its true value. The huge mass and unphysical mass obtained is then explained by stating that there is a supermassive Black Hole at the center of the galaxy. Thus, the belief that there is a Black Hole, at the center of each galaxy, lies in the fact that current astronomy and astrophysics, in spite of overwhelming evidence, refuses to accept that redshift is not due to recessional velocity, but that it is intrinsic, instead.

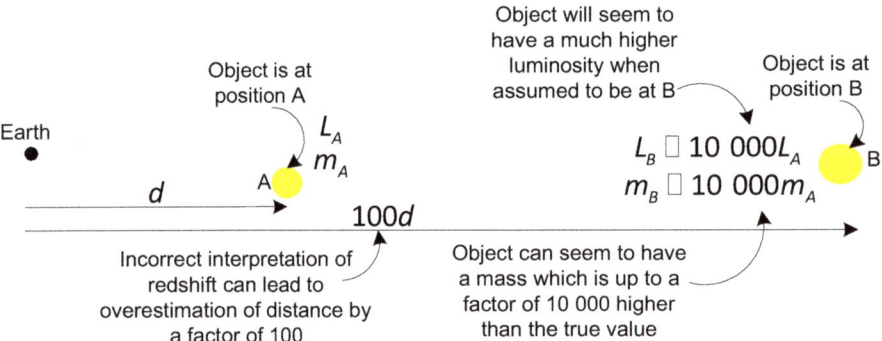

Figure 30.8. The interpretation of redshift, as recessional velocity, and therefore distance, can lead to an overestimation of distance, which then leads to a huge overestimation of luminosity and therefore mass.

Figure 30.9. The Seyfert galaxy NGC4258, with a redshift of 0.002, and known to eject material from its active nucleus has two quasars of redshift 0.4 and 0.65 on either side of the nucleus [1]

The image shown in figure 9, and other images discussed by Arp in 'Seeing Red' lead to the conclusion that galaxies with active galactic nuclei, in other words galaxies which have a small central region, which is much brighter, or luminous, than the rest of the galaxy, or are releasing substantial amounts of energy, periodically eject material, from the nucleus, which then condenses into quasars. The ejection is along the galaxy's minor axis (see figure 10) and in opposite directions, and thus leads to the appearance of quasars, on opposite sides of the galactic nucleus. The newly ejected material has an extremely high redshift but the redshift decreases as it ages. As the quasars age, they also eject material from their centers, and this material turns into the arms of a companion galaxy. It is called companion because it will stay close to the parent Seyfert galaxy and will not drift far from that galaxy's minor or rotational axis. The fact that so much material is ejected from the nuclei of galaxies suggests that instead of Black Holes at their centers, there are White Holes. White Holes continuously eject matter instead of absorbing it. It therefore seems matter is continuously being created at the center of galaxies.

Figure 30.10. On the left: Active Galactic Nuclei are intensely bright galactic centers, with extremely high electric fields. When the electric field and brightness reaches a critical level, it causes instability and the ejection of matter, which condenses into a quasar. In the course of time, the quasars start ejecting their own material, along their major axis, which develops into arms, and thus become galaxies themselves. The material ejected from the center of quasars spreads out in a spiral because of the rotational motion of the quasar. Thus galaxies give birth to galaxies, and matter is continuously being created in the universe. On the right: A spiral galaxy, the spiral arms are due to material being ejected along the galaxy's plane of rotation.

This matter is not created out of nothing though, it seems that the galaxies become extremely bright before an ejection event, which suggests that the intense electromagnetic fields, it generates may lead to the creation event. In other words, there is an energy transformation that leads to the creation of matter so that electromagnetic energy is transformed into ejected matter. Initially, this matter has close to zero mass, and can thus move close to the speed of light, but as it ages, its mass increases and its speed decreases. This can be seen from the fact that a highly redshifted object will have a low frequency and therefore low energy, which is equivalent to low mass.

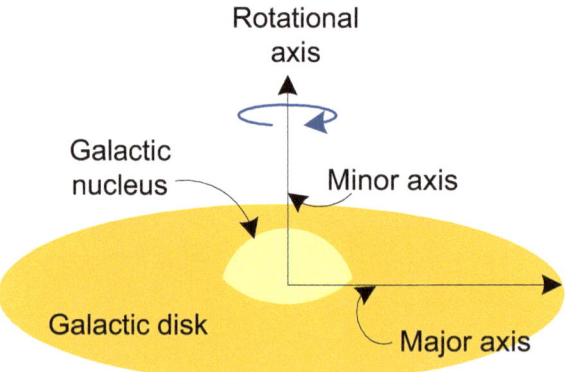

Figure 30.11. A galaxy's major axis is along its plane of rotation, or along the galactic disk, and the minor axis is perpendicular to the galactic disk, or along the axis of rotation.

The reason why quasars and therefore companion galaxies, which form from material ejected from a parent galaxy, with a much lower redshift, do not move very far away from the direction, from which they were ejected is because they are ejected along the axis of rotation of the parent galaxy, and thus tend to continue to move from the parent along that same direction.

In conclusion, instead of a Black Hole at the center of galaxies, we have continuous matter creation, or what can be termed a White Hole. Because redshift has been shown not to be due to recessional velocity, but rather, that it is intrinsic, the big Bang model has been falsified. Thus, instead of one creation out of nothing event, leading to an unstable universe (Big Bang), we have continuous matter creation out of electromagnetic energy. The fact that current astrophysicists and astronomers refuse to accept the truth has led to just about everything being reversed in this field. Thus, instead of matter creation, Black Holes gobbling up matter are envisioned. Instead of galaxies forming as a result of matter ejection, galaxies are perceived to be colliding. Instead of outward motion of matter, everything is perceived to be collapsing inward.

References:

[1] Arp, Halton (1998). *Seeing Red*. Apeiron, Montreal.
[2] Albers, C. (2017). Article 103: The 800 million solar masses black hole at the edge of the universe.

The End for Now!

www.ingramcontent.com/pod-product-compliance
Lightning Source LLC
Chambersburg PA
CBHW051912210526
45473CB00006B/1979